Fermentation

a practical approach

$1\cancel{4}$

TITLES PUBLISHED IN
THE
PRACTICAL APPROACH
SERIES

Series editors:
Dr D Rickwood
Department of Biology, University of Essex
Wivenhoe Park, Colchester, Essex CO4 3SQ, UK
Dr B D Hames
Department of Biochemistry, University of Leeds
Leeds LS2 9JT, UK

Affinity chromatography
Animal cell culture
Antibodies I & II
Biochemical toxicology
Biological membranes
Carbohydrate analysis
Cell growth and division
Centrifugation (2nd Edition)
Computers in microbiology
DNA cloning I, II & III
Drosophila
Electron microscopy
in molecular biology
Fermentation
Gel electrophoresis of nucleic acids
Gel electrophoresis of proteins
Genome analysis
HPLC of small molecules
HPLC of macromolecules
Human cytogenetics
Human genetic diseases
Immobilised cells and enzymes
Iodinated density gradient media
Light microscopy in biology
Liposomes
Lymphocytes
Lymphokines and interferons
Mammalian development
Medical bacteriology
Medical mycology
Microcomputers in biology

Microcomputers in physiology
Mitochondria
Mutagenicity testing
Neurochemistry
Nucleic acid and
protein sequence analysis
Nucleic acid hybridisation
Nucleic acids sequencing
Oligonucleotide synthesis
Photosynthesis:
energy transduction
Plant cell culture
Plant molecular biology
Plasmids
Prostaglandins
and related substances
Protein function
Protein purification applications
Protein purification methods
Protein sequencing
Protein structure
Proteolytic enzymes
Ribosomes and protein synthesis
Solid phase peptide synthesis
Spectrophotometry
and spectrofluorimetry
Steroid hormones
Teratocarcinomas
and embryonic stem cells
Transcription and translation
Virology
Yeast

Fermentation

a practical approach

Edited by
B McNeil and L M Harvey

Division of Applied Microbiology,
Department of Bioscience and Biotechnology,
University of Strathclyde,
204 George Street,
Glasgow G1 1XW, UK

OXFORD UNIVERSITY PRESS
Oxford New York Tokyo

Oxford University Press
Walton Street, Oxford OX2 6DP

British Library Cataloguing in Publication Data
  Fermentation.
  1. Fermentation
  I. McNeil, B. II. Harvey, L.M.
    547'.29

Library of Congress Cataloging in Publication Data
  Fermentation: a practical approach/edited by B. McNeil and L.M. Harvey.
  Includes bibliographical references.
  1. Fermentation. I. McNeil, B. II. Harvey, L.M.
    TP156.F4F43 1989   660'.28449—dc20   89-19966
  ISBN 0 19 963044 5
  ISBN 0 19 963045 3 (pbk.)

Previously announced as:
ISBN 1 85221 155 5
ISBN 1 85221 156 3 (pbk.)

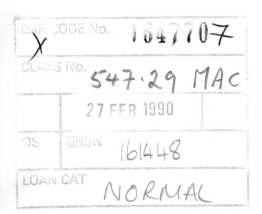
Typeset and printed by Information Press Ltd, Oxford, England

Preface

Fermentation is a term which is interpreted in greatly differing ways by those involved in the art; for example, the brewer would have quite a different concept of fermentation from that of the penicillin producer. In this book, we have concentrated largely on the selection and use of laboratory scale stirred tank fermenters (or bioreactors) for the cultivation of microbial and animal cells. It soon became clear however, that there were many other factors which had to be taken into account in order to run a fermentation. It was therefore felt necessary to include chapters on laboratory and inoculum preparation, fermentation modelling, sterilization, and instrumentation and control.

Regardless of their original subject, newcomers to fermentation face two clear difficulties: first, they must quickly understand something of a wide range of disciplines; secondly, they must put that knowledge to practical use in the operation of the selected fermentation process. Newcomers will learn as all newcomers in the past have learnt, by reading, by discussion with those more experienced than themselves and, of course, by making mistakes! Most books on the subject of fermentation technology are process-oriented, often theory-centred and tend to give more complete coverage to industrial scale than laboratory scale operations. The emphasis in this book is deliberately upon the laboratory scale practical applications with only a minimal coverage of the relevant theory. The aim of this book is to pass on some of the practical knowledge accumulated by those who have worked in the area of laboratory fermentation for a number of years. It is our hope that it will impart some of the knowledge common to fermentation technologists which to date has rarely been collected together in written form.

B.McNeil and L.M.Harvey

Contributors

A.Brown
Delta Biotechnology, 5 Crocus Street, The Meadows, Nottingham NG2 3DE, UK

D.Cantero
University of Cadiz, Cadiz, Spain

M.K.Dawson
School of Chemical Engineering, University of Birmingham, PO Box 363, Birmingham B15 2TT, UK

P.J.B.Dusseljee
Gist Brocade, Wateringsweg 1, Delft 2600 MA, The Netherlands

J.Feijen
Gist Brocade, Wateringsweg 1, Delft 2600 MA, The Netherlands

P.J.Halling
Applied Microbiology Division, University of Strathclyde, 204 George Street, Glasgow G1 1XW, UK

L.M.Harvey
Applied Microbiology Division, University of Strathclyde, 204 George Street, Glasgow G1 1XW, UK

D.C.Hearle
School of Chemical Engineering, University of Birmingham, PO Box 363, Birmingham B15 2TT, UK

T.S.Irvine
Beecham Pharmaceuticals, Shewalton Road, Irvine, UK

C.A.Kent
School of Chemical Engineering, University of Birmingham, PO Box 363, Birmingham B15 2TT, UK

M.Lavery
Hoechst Animal Health, Walton Manor Walton, Milton Keynes MK7 7AJ, UK

B.McNeil
Applied Microbiology Division, University of Strathclyde, 204 George Street, Glasgow G1 1XW, UK

C.Parton
ICI, Alderly Edge, Merside, Macclesfield, Cheshire SK10 4TG, UK

C.G.Sinclair
Department of Chemical Engineering, UMIST, PO Box 88, Manchester M60 1QD, UK

D.J.Weale
School of Chemical Engineering, University of Birmingham, PO Box 363, Birmingham B15 2TT, UK

P.Willis
Biological Products Division, ICI, PO Box 1, Billingham, Cleveland TS23 1LD, UK

Contents

5. FED-BATCH AND CONTINUOUS CULTURE
A.Brown

6. pH, DISSOLVED OXYGEN AND RELATED SENSORS
P.J.Halling

9. ANIMAL CELL FERMENTATION
M.Lavery

APPENDIX

INDEX

CHAPTER 1

Setting up – equipment for laboratory fermentation

BRIAN MCNEIL and LINDA M.HARVEY

1. INTRODUCTION

When considering the setting up of a laboratory for fermentation, a number of points should be kept in mind.

(i) *Scale of operation.* Clearly, at the bench-top scale (up to 3 litres volume), little additional adaptation would be required in a standard microbiology laboratory. Many of the items of ancillary equipment, such as tubing, clamps, pumps, and shakers, are common to all scales of operation. Larger-volume fermenters (usually mounted on skids, trolleys, or frames) require more extensive preplanning of the operational area. Such fermenters, from 5 – 50 litres working volume, might well be termed research scale; anything larger would come into the category of a small pilot plant. Research- or laboratory-scale fermenters will be covered in more depth in Chapter 2.

(ii) *Type of fermenters.* For particular applications (outlined in Chapter 2) specialized fermenter types such as air-lift or loop reactors may have their uses, but for most purposes the stirred tank bioreactor is best. It is a well-characterized type of reactor, used at every operational scale, and has a much higher degree of operational flexibility than most other fermenters.

(iii) *Number of fermenters.* The use of one or two bench-top fermenters will require only a source of electrical power, water, and access to a small autoclave. Operation of a number of larger vessels will have considerable impact on lab design; for example, it becomes worthwhile arranging steam lines to each fermenter point to allow for *in situ* sterilization.

(iv) *Nature of fermentation process.* This really relates to the type of organism to be used; the use of animal cells will require equipment quite different in some respects from that employed for 'traditional' microbial fermentation. This is covered in Chapter 9.

In the description of a fermentation laboratory which follows, it should be remembered that this describes the 'ideal', i.e. a lab constructed *ab initio* for the purposes of microbial fermentation. Such an 'ideal' is rarely achieved, but is always a useful target.

2. THE LABORATORY

2.1 General description

A fermentation laboratory is primarily a microbiology laboratory incorporating advanced technology equipment, and as such should be designed to maximize efficiency while maintaining a high safety standard. To this effect the laboratory should be well designed to allow research to progress in a comfortable environment, with well-planned operating areas. The general layout of the laboratory will depend on the shape and size of the room allocated, and it is advisable for laboratory personnel to liaise closely with architects and engineers to produce a design that is economically feasible while still satisfying the requirements of research personnel.

The situation of the laboratory is important, because fermentation laboratories house heavy pieces of equipment and the floor of the laboratory must be able to take heavy loads. In addition, delivery of heavy equipment and bulk chemicals requires ease of access to the laboratory.

The laboratory area should be light, spacious, and well ventilated. The latter is of particular importance, because fermentation laboratories can become humid and unpleasant to work in. It may, therefore, be advisable to improve the control of temperature and air quality by supplying an air-conditioning unit.

Space is required not only for a wide variety of general laboratory practices, but also for cleaning, stripping down, and repairing fermenters. The laboratory should also contain sufficient cupboards and shelves to house general equipment such as glassware, laboratory plastics, etc., and also fermenter accessories. It is advisable to allocate one cupboard per fermenter for spares and accessories that are specific to that fermenter. A central store for spares common to all fermenters, for example, O-rings, probes, connectors, etc., is very useful. All laboratory furniture must be easily cleaned and solvent resistant, and should not be affected by decontaminating solutions.

The laboratory is divided into two or three distinct areas:

(i) *Wet floor area.* This houses all fermenters, stills, autoclaves.
(ii) *General laboratory area.* Used for medium preparation and basic 'wet' processes, i.e. dry weights and viscosity measurements.
(iii) *Analytical area.* Used for analysis of the fermentation broths. This area will contain, in a well-equipped laboratory, items of equipment such as GLC, HPLC, spectrophotometers, etc.

The analytical section of the laboratory should be as far away from the wet floor area as possible, to prevent damage to sensitive analytical equipment. *If possible*, all analytical equipment should be housed in a separate but adjacent laboratory. Humidity generated in the fermentation area can seriously affect analytical equipment resulting in the production of inaccurate results or, in extreme cases, complete equipment failure. Vibration produced by the fermenters, and ancillary equipment such as centrifuges, can also cause inaccurate readings. This is especially evident where balances are concerned. If a separate room cannot be provided, balances must be housed on a stone bench as far away from the fermentation area as possible in order to reduce the vibration effects.

2.2 **Floor areas**

A fermentation laboratory has two distinct areas, the wet floor and the dry floor.

2.2.1 *Wet floor*

All fermenters, including bench-top varieties, should be situated on the wet floor area so that spillages, intentional or not, can do no damage either in the laboratory or to people and equipment situated on the floor below. Wet floors should be constructed to meet two main criteria. They should be non-slip, whether wet or dry, and should be easily cleaned. These factors are often not compatible, i.e. the rough surface required to make the floor non-slip is often ideal for harbouring dirt. Specialist floor manufacturers are available in most large towns and cities, and should be consulted for information on the latest flooring materials available. The wet floor should be constructed so that it slants towards a central drainage channel (fermenters should be levelled using adjustable feet or blocks); the channel itself should slant towards the drainage output of the laboratory, and must be covered by a stainless steel or heavy-duty plastic grid to prevent accidents.

2.2.2 *Dry floor*

The dry floor area of the laboratory should be covered with seamless sheet vinyl which has a smooth surface and does not catch dirt.

The junction between the wet and dry floor areas of the laboratory is an area of potential danger, as sheet vinyl is slippy when wet. It is advisable to introduce an area where shoes can be dried between the two floor types.

Routine cleaning of both floor areas is essential. A wide range of detergents/disinfectants is available, and the product chosen will depend on the biocidal effectiveness required. Because of the nature of equipment and fermenter contents, trained personnel only should clean the wet floor. If the wet floor area is large, fire hoses are very useful for cleaning and rinsing purposes.

3. SERVICES

A number of essential services are required to run an efficient fermentation laboratory, namely air, steam, water, electricity, and gas. The level of sophistication of each service required depends on the size of laboratory and the demands on that service. All services should be available 24 hours a day, seven days a week, for a dedicated fermentation laboratory; smaller laboratories housing only a small number of simple fermenters do not require this level of service.

3.1 **Air**

Air is required for many purposes in a fermentation laboratory, e.g. aeration of fermenters, operation of hydraulic autoclave doors, and calibration of gas analysis equipment.

Laboratory fermenters are often supplied with integral air pumps. Individual air pumps can be purchased to supply air vessels which do not have an integral air supply and which are situated in laboratories without service air. The ideal pump is the diaphragm pump. These pumps vary greatly in their capacity and should be purchased according to the demand of the vessel. It is essential that the pumps be oil free and be suitable

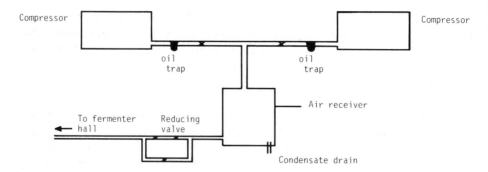

Figure 1. Arrangement of air supply system for a fermentation laboratory.

for prolonged continuous usage. Diaphragm pumps are relatively cheap and are useful if only one or two vessels require an air supply.

For larger laboratories, a more sophisticated air supply is required and a compressor should be purchased. The type and size of compressor selected depends on the function it is to have in the laboratory. The compressor should be able to meet the current demands of the laboratory with additional capacity to allow for subsequent growth.

Fermentation processes require particle-free clean air that meets food grade standards, so the compressors utilized must be oil-free units. Beware of compressors which are claimed to produce oil-free air; these are not oil-free compressors and should be avoided since even when functioning well they can allow oil to escape into the system, leading to fouling of lines and contamination of the fermentation. If such a system is already installed, the air lines must have reliable protective filters downstream of the compressor which are regularly cleaned and serviced.

To ensure a constant supply of air at all times it is advisable to operate two compressors so that if one fails for any reason, the other unit will cut in and maintain the supply going. It is necessary to ensure that each compressor is used on alternate days so that the demand on each machine is equal. Incorporating an air receiver in the supply line (*Figure 1*) decreases the demand on the compressors and also allows condensate to be removed from the supply. Provision of two compressors also allows maintenance procedures to be carried out without interrupting the supply to the laboratory.

If the compressor is to be situated in the laboratory, acoustic hoods should be fitted to reduce the noise level. This is not an expensive exercise, and certainly allows laboratory personnel to work in greater comfort. Ideally the compressor(s) will be housed in a room situated adjacent to the fermentation laboratory.

The air supply is normally taken from the compressor to the fermenter via a network of pipes. The pipework should be made of a non-corrosive material with a smooth interior to prevent build-up of dirt. Half-inch steel piping is excellent for this purpose and is also relatively cheap. The air supply from the compressor is normally delivered to the laboratory at a much higher pressure than is required. Pressure-reducing valves must, therefore, be installed downstream of the compressor along with oil and water traps (as compressed air expands in the lines a small amount of water is formed—0.025 g H_2O per litre air at 25°C). All reducers and traps should be regularly cleaned and serviced. All connections made in the lines must be able to withstand the pressure at which the air is to be delivered (frequently $1-2$ bar g, but this depends on the fermenter(s) used). It is, therefore, necessary to have the system of pipework installed by a qualified plumber.

3.2 Steam

Steam is required in a fermentation laboratory for sterilizing fermenters, controlling temperature in larger vessels, autoclaving, and supplying steamers. Each fermenter bay should have its own steam supply which can be isolated from the vessel by means of a gate valve. Supply 24 hours a day is required only if fermenters using steam for temperature control are being used, or if there is a heavy load on the autoclave, requiring runs during 'out of work' hours. It is essential, however, to have steam supplied to the laboratory seven days a week. If the building already has an efficient steam supply, then all that is necessary is to tap into the system. If a steam supply is not available then it may be necessary to go to the expense of purchasing a steam generating plant. Such plants range in complexity from simple boilers with small capacity to highly sophisticated units.

The steam supply should be as dry as possible, and all lines in the laboratory should be well lagged to help prevent the formation of condensate. Lagging is also important for the protection of laboratory personnel. The lagging should be covered with an aluminium casing to give a finish that looks more presentable, is easily cleaned and prevents the lagging material getting damp.

As with the air supply it may be necessary to incorporate pressure reducers in the lines. These should be regularly inspected and cleaned. Installation of the steam lines should be carried out by qualified personnel.

3.3 Water

A fermentation laboratory requires a constant supply of water to service fermenters, downstream processing equipment, and autoclaves, as well as for analytical purposes.

Mains water is normally connected to the service inlet on the fermenter chassis. Manufacturers will indicate the maximum inlet pressure a particular model of fermenter can cope with. This is often very different for different vessels, e.g. a small lab fermenter is generally serviced by water supplied at 1 bar g whereas a small pilot-scale vessel may require up to 3 bar g. The mains water inlet pressure must be sufficient to meet the demands of the laboratory; if necessary, pressure reduction valves can be fitted on line to each vessel to allow a range of different demands to be serviced.

Waste water from the cooling/condenser systems must pass to a drain via the waste water outlet line on the chassis. Robust polyester-reinforced PVC or pressure tubing of any kind can be used for this purpose; alternatively, the waste water can be directed to the drain via a fixed, usually copper, line.

The laboratory must also be supplied with good-quality water, for medium preparation and analytical work. There are many types of still available for the production of distilled water, the size of still required being determined by the demand of the laboratory. Where ultra-high-quality water is required, e.g. for HPLC, ion chromatography, etc., it may be necessary to install a deionizing system, particularly in areas where the local supply water is hard. There is a wide variety of models to choose from, each system incorporating reverse osmosis, adsorption, and deionizing units for the production of ultra pure water. It should be noted that such systems are expensive to run because the cartridges in each unit must be replaced at regular intervals.

Water is also required for the basic laboratory chores of washing glassware, etc., and hand washing. As well as ordinary sinks, it is essential to include a waste disposal unit in the fermentation laboratory, particularly if the laboratory is handling mycelial cultures which tend to block sinks if they are not previously macerated. Small bench sinks and taps are useful for many purposes, including vacuum filtration.

3.4 Electricity

Electricity is required for lighting and for supplying power to the complex array of machinery and instrumentation found in a laboratory. Lighting should be moderately priced, have diffusers with a high power factor, and be as noise-free as possible. The power supply to the laboratory should meet the demands of all equipment therein and allow capacity for future growth.

Many items of equipment in a fermentation laboratory draw large amounts of power, e.g. small pilot-scale vessels, autoclaves, and downstream processing equipment. It is, therefore, necessary in dedicated fermentation laboratories to have a supply of three-phase electricity. Installation of three-phase supplies can be expensive; it is advisable, therefore, to have a supply in the laboratory and to take off lines as and when required.

Every fermenter in the laboratory will require a number of electrical sockets supplying power to the fermenter, pumps and additional equipment, such as portable pH and dO_2 meters. A bank of six sockets per fermenter is ideal for this purpose. In a laboratory that has not been specifically designed for fermentation, it is possible to overcome a lack of power points by the use of socket boards. Care should be taken not to overload the electrical capacity of the laboratory if this system is adopted. For larger fermenters requiring a three-phase supply it is still very useful to have a bank of single-phase sockets which should be installed alongside the three-phase supply.

3.5 Gas

Natural gas should be supplied to the general laboratory area for general laboratory usage. Speciality gases, e.g. carbon dioxide, oxygen, and calibration gases, are usually provided in the form of compressed gas cylinders which should be secured with safety straps. If large quantities of such gases are required, consideration should be given to setting up a permanent rack outwith the laboratory and piping the gases in to service outlets within the laboratory.

4. FERMENTERS

A well-equipped fermentation laboratory requires substantial capital investment for the purchase of fermenters and a wide range of ancillary equipment. This chapter deals with the basic equipment required to start up a fermentation laboratory and makes no reference to the analytical equipment necessary for detailed analytical work, i.e. gas analysis equipment, HPLC, and GLC.

The wide range of fermenters available and their advantages and drawbacks are discussed in Chapter 2. A fermenter can be free-standing on its own frame, skid-mounted, or bench top, and should be housed on the wet floor area of the laboratory. The wet floor area can be divided into a number of fermenter 'bays', with sufficient room between each vessel to allow easy operator access. Each bay will contain all the services required by the fermenter, i.e. six electric sockets, two water supplies, an air supply, a steam supply and an off-gas analysis point. It may be necessary to colour code the service lines to allow easy identification and avoid confusion.

If the fermenter is free-standing, check the area available on the mainframe for additional equipment, and for pumps, acid and alkali reservoirs, etc. If the area available is too small, wall space must be available for shelving; alternatively, there should be room for a small bench to be located next to the fermenter.

5. ANCILLARY EQUIPMENT

5.1 Autoclaves

A wide range of autoclaves is available for use in the laboratory. The size and type of autoclave purchased will depend on the number of fermenters being serviced. Frequently, in the fermentation laboratory, it is useful, but not essential, to have more than one type of autoclave. Many fermenters of 10 litre volume and above can be purchased with their own sterilizing facility so that the vessel and contents can be steriliz-ed *in situ*. If all the fermenters in the laboratory have this facility a standard-sized autoclave, e.g. 450 × 800 × 1200 mm chamber dimension, will be sufficient to deal with ancillary purposes, e.g. sterilizing glassware, or preparation of small quantities of media.

5.1.1 *Bench-top or portable autoclaves*

Bench-top or portable autoclaves are useful for sterilizing small items, i.e. bottles, a small number of 500 ml conical flasks, automatic pipette tips, etc. Such autoclaves work on the same principle as a household pressure cooker. Pressure cookers can be used for this purpose, but the lifespan of the portable autoclave is generally longer, because it is more robust in construction. Programmable portable autoclaves are available but these are about six times as expensive as the standard portable autoclave. The standard portable unit usually requires a form of external heating, although electrically heated models are available, and a gas burner should be purchased to supply that heat. Standard models should not be left unattended for any length of time, as they may boil dry and damage both the autoclave and its contents. Portable autoclaves are generally made from diecast aluminium alloy or stainless steel, and the inner chamber should be easy to remove to facilitate cleaning. Regular checks should be made on the pressure and safety valves to ensure that they are clean and functional.

5.1.2 *Standard autoclaves (fast turnround)*

Small autoclaves are useful for sterilizing small volumes of medium (up to 10 litres), general laboratory equipment, e.g. pipettes, and also vessels from 1 to 10 litres capacity which cannot be sterilized *in situ*. If the autoclave is being used for this purpose, the height of the autoclave chamber must be sufficient for the vessel plus support stand

7

or legs plus any protrusion from the top plate of the vessel, e.g. probes or sensors.

Standard autoclaves either require an independent steam supply or are of the electric type, working on the principle of a small boiler. When using an electric autoclave, care must be taken to ensure that the elements are always covered with water to prevent them burning out. Electric autoclaves are much slower than direct steam models because time is required to heat the water to boiling point. This is an important factor to consider if a fast turnround of material is required.

This size of autoclave comes in two designs, vertical or top loaders and front loaders. The bulk of material required to be sterilized in a fermentation laboratory is generally heavy and it is, therefore, impractical and possibly dangerous for laboratory personnel to use a top-loading machine. If a top-loading machine is all that is available, then a hoist should be used to load and unload the chamber.

5.1.3 *Large autoclaves*

In a large facility or a small laboratory with a number of fermenters without an *in situ* sterilizing programme, a large autoclave is required to cope with the demand for sterilizing vessels and reservoirs full of medium. Large autoclaves are particularly useful in laboratories that practice a lot of continuous culture where the demand for large quantities of media is high. Large autoclaves require an independent direct steam supply and a three-phase power supply. These factors should be taken into consideration when purchasing the autoclave, as they can increase costs considerably. Large autoclaves are front-loading models and it is advisable to purchase a loading platform and trolley to aid the loading of the chamber.

The cycle time for a large autoclave is usually long, i.e. 3−5 hours for a full load. This should be borne in mind when calculating the throughput possible in a 24-hour period, in order to determine whether the machine will meet the demand of the laboratory.

5.2 Incubators, orbital incubators, and shakers

Incubators are required for the cultivation of stock cultures and production of inocula. They range in size from controlled-temperature rooms to floor-standing units of varying size. It is useful to have at least two incubators in the laboratory so that a range of organisms can be grown at different temperatures. If the organism is required to grow at temperatures below 25°C it may be necessary to employ a cooling or refrigeration unit with the incubator. Incubators should be regularly cleared out and swabbed with a suitable disinfectant.

Orbital incubators and shakers are required for the production of inocula, where a spore inoculum is not being used, and also for shake-flask experiments, e.g. for the purposes of basic medium design. The orbital units are designed so that the motion of the platform brings about mixing of flask contents and gas transfer, and aims to minimize wall growth on the side of the conical flasks. Nevertheless wall growth is a common problem, particularly in the cultivation of filamentous organisms. Orbital incubators are more expensive than orbital shakers but are much more versatile, allowing a wide range of temperatures to be used. Orbital shakers must be used in a controlled-temperature room and are, therefore, less flexible than orbital incubators, because operating temperatures are restricted by the availability of controlled-temperature rooms. It may be useful to buy more than one platform for the incubator/shaker so that different sized flasks can be used. In some cases it is possible to alter the flask size held by adjusting the spring clips on the platform. In order to maximize the flexibility of the equipment in the fermentation laboratory, variable-speed orbital incubators/shakers should be purchased.

Platforms are usually covered in plastic laminate to facilitate easy cleaning. All spillages should be cleaned up as soon as possible; fermentation media generally contain a high concentration of sugar which can affect the smooth running of the shaker if allowed to dry on to the drive belt. If breakages occur, care should be taken to remove any glass fragments. The spring clips on the platform should be regularly checked, and damaged clips removed and replaced. These three simple precautions should help to extend the life span of the equipment considerably.

5.3 Ovens

Two types of oven are useful in fermentation:

(i) *Hot air ovens* can be used for drying glassware or dry weights, or as sterilizing ovens. The working chamber should be made of corrosion-resistant stainless steel. Most hot air ovens rely on good air circulation; air is drawn into the oven, passes over a heating element and then passes out. It is therefore, important to site the oven in an area where good circulation of air is available, i.e. not against a wall or other piece of equipment.

(ii) *Microwave ovens* are used for dry weights, drying glassware, melting agar, etc. The main advantage of employing a microwave is that the above processes can be carried out very quickly, i.e. 20 minutes to do a dry weight (cf. overnight in a hot air oven). Thus, quick determinations can be obtained, allowing daily assessment of the fermentation. Dry weights must be carried out using the defrost setting of the microwave. Using a higher setting can damage the oven and tends to result in scorching. Agar can be melted on either a medium or high setting, depending on the model of oven employed. It is worthwhile spending time determining appropriate settings and times for each process, rather than using a trial-and-error approach each time the oven is used. Old or thick glassware may shatter if used in a microwave, but modern laboratory glassware, i.e. Pyrex or Schott, can be used safely. Manufacturers' instructions should be carefully followed, i.e. no metal objects, such as bottle caps, should be put in the microwave. Any queries will usually be answered by the manufacturers' service department.

5.4 Pumps

At laboratory scale, liquid pumping is achieved almost exclusively by means of peristaltic pumps; diaphragm dosing pumps (discussed in Chapter 5) also find some application but are less flexible. From the fermentation technologists' viewpoint, such a peristaltic pump has a number of clear advantages:

(i)   The fluid contained within tubing does not contaminate the pump;
(ii)   The pumps are self-priming;
(iii)   No back-flow occurs even when the pump is off (i.e. no check valves are needed);

(iv)   This means of pumping is relatively mild, which may be important with some microbial cultures;

(v)   Flow rate is very consistent.

    The choice of pump will depend on the application; we commence with the smallest.

5.4.1 *Minipump (or delta pump)*

These fixed-speed pumps are normally used for addition of acid/alkali, antifoam linked to a pH/antifoam controller. Such pumps, even fitted with the widest bore tubing (5 mm), have an output of at most 25 ml min^{-1}, which is adequate for most titrant addition functions in laboratory fermenters. They are reputed to be capable of pumping against pressures of up to 5 p.s.i. (0.34 bar) though this depends on a number of factors including tube tension. When not in use, the pump tubing should be taken off the pump head to avoid sealing of the tubing (especially with acid/alkali lines). A smear of silicone grease on either the rollers or the pump tubing will reduce wear considerably. Since the bore of tubing is finer than that for general use, it is advisable to order a supply of tubing (silicone) and pump nipples.

5.4.2 *Larger pumps*

Larger pumps, used for example for nutrient addition in continuous or fed-batch culture, generally have clamps to hold tubing on the pump heads. Such pumps, which take tubing of up to 8 mm bore, can be either fixed speed or variable speed, manual or auto control. By means of changing the bore of tubing used, even the output of a fixed-speed pump can be varied. However, in general, it is worth paying for the extra flexibility given by the variable-speed type pumps, which may cost up to twice the price of the equivalent fixed-speed model. As with the minipumps, the pump heads should be lightly greased before use, and tubing removed from the pump head when not in use. In order to even out wear (the tubing section in the pump tends to stretch with time), the section of tubing in the pump head should be changed. Unless specifically required, remote-controlled pumps are not generally worth purchasing; the manual control version is cheaper and adequate for most work.

    It is often necessary to sterilize larger volumes of medium separately from the fermentation vessel, then pump the sterile medium into the pre-sterilized vessel. For such purposes, a high-speed pump, capable of delivering at least $3-5$ litre min^{-1} using standard $5-6$ mm bore tubing, is very useful. Most manufacturers supply such pumps.

    When buying a pump, it is important to think carefully about the intended flow rate range. Due to the nature of electric motors (and of the tubing), pumps are most accurate at high speeds and least accurate at very low settings. On the other hand, tube life is much reduced at high speeds. Since at laboratory scale tube life is rarely as much of a consideration as experimental accuracy, it is best to choose a pump which can produce the desired output when operating in mid range settings.

5.5 **Pump tubing**

Table 1 summarizes the characteristics of certain tubing types suitable for use with peristaltic pumps. Since the tubing for use with peristaltic pumps is often quite different from 'general purpose' tubing, the characteristics may differ significantly.

Table 1. Tubing types for peristaltic pumps (Courtesy of Watson-Marlow Ltd).

Type	Autoclavable	Chemical sterilization (ethylene oxide)	Gas permeability	Temperature limit* (°C)	Transparent/ translucent	Cost relative to silicone
Silicone	yes	no	high	135	translucent	1.0
Marprene	yes	yes	low	135	no	2.0
Neoprene	no	no	medium	80	no	1.1
Butyl	yes	no	v.low	80	no	1.2
PVC	no	yes	v.low	60	transparent	0.7

*This refers to the temperature of liquids which can be pumped using tubing of each type.

The most useful tubing for all purposes is silicone which is flexible, durable, non-toxic, autoclavable, and can be visually inspected for contamination, etc. Marprene (Watson-Marlow Ltd), and similar products of other companies, are specifically designed as pump tubing and have a greatly extended lifespan (up to 10 times) compared to silicone in this role. At the lower end of the lab scale, the extra cost of equipping pumps with Marprene or equivalent is not justified, and silicone is a satisfactory substitute. Another potential drawback of Marprene is its opaque nature, which impedes visual examination. Butyl tubing has one major advantage over silicone in that it is far less permeable to gases; it is therefore recommended for cultivation of anaerobes, where oxygen penetration of medium and culture lines is best avoided.

PVC, especially nylon/polyester reinforced PVC, has a number of roles in the laboratory where a more rigid, less easily compressed tubing is essential, e.g. water lines to and from the fermenter gas pressure lines.

Silicone is, therefore, the tubing in most frequent use, both for pumps and for general purposes, with bores of 5−8 mm being the most common sizes.

5.6 Connectors and clamps

The use of flexible silicone tubing considerably eases the making of junctions; in many cases, with a suitable hose tail fitting the tubing can simply be pushed on. The disadvantage is, of course, that it is equally easy for tubing to slip off, or be pulled off, by the action of a pump, for example. In order to avoid this, it is helpful to acquire a stock of various clips, which should include worm drive, hexagonal-headed hose clips, and the much cheaper acetyl plastic snapper type. Nylon cable ties are also useful. Clamps, for sealing sections of tubing, are also essential, and a considerable number will be required. The type made of steel with a finger screw adjustment and crossbar are most useful.

The easiest means of attaching one piece of flexible tubing to another (or several others) is by use of tubing adaptors. Where they must undergo autoclaving, polypropylene connectors should be used. It is advisable to have a range of sizes available which should include T and Y piece types.

5.7 Filters

The commonest type of filtration process carried out in fermentation is the removal of particles (microorganisms, dust, etc.), from process air or gas. Most fermenters will therefore have a gas filter on the inlet gas line, and, as mentioned in the section on safety, a similar filter on the exit gas line. Such filters can be of several types.

5.7.1 *Membrane type filters*

These have filter holding containing a cellulose acetate, or nitrate, or PTFE membrane of known and consistent pore size, which therefore retains all particles larger than that pore size. 0.2 μm or 0.45 μm pore size filters are suitable for most applications. A filter with a suitable flow range should always be selected. These filters (commonly described as miniature in line), are relatively cheap, disposable, and usually readily inspected for blocking or fouling. They should be discarded after a fixed number of autoclave cycles. The manufacturer will normally indicate the number of cycles which can be withstood.

5.7.2 *Packed-bed type filters (depth filters)*

These filters have no uniform pore size and the mechanism of particle removal tends to be rather more complex. Typically, a filter housing is packed with glass wool or non-absorbent cotton wool. Such filters are vulnerable to compaction, and to wetting which may allow channelling to occur. Sudden fluctuations in the pressure drop across the filter can cause release of particles or packing material under some circumstances. Such filters are readily constructed in-house, but their sole advantage is cheapness.

5.7.3 *Cartridge filters*

These are composed of a stainless steel or polycarbonate filter housing containing a removable filter element. The filter element is often composed of a hydrophobic material (e.g. PTFE) bonded in polypropylene. (Hydrophobic filters tend to be less susceptible to degradation of filter efficiency in wet conditions.) These filters can be steam sterilized *in situ* or autoclaved. They are initially more expensive than other types, but their reliability and durability make up for this.

5.8 **Medium reservoirs**

Particularly in fed-batch and continuous culture (Chapter 5), it is essential to have a number of medium supply vessels. Such vessels must be able to withstand repeated autoclaving. The volume of vessel required (and the number) will depend on the experimental programme, e.g. a 10 litre fermenter operated at a dilution rate of 0.1 h^{-1} will use 24 litres per day, so 5 litre reservoirs would be inappropriate. In practice, glass (heat-resistant) and polycarbonate vessels are utilized. These are obtainable up to 20 litres total volume. Smaller bottles of 10 litre, 5 litre, and less, are also available. At the 20 litre scale, polycarbonate vessels are superior to glass because they are lighter, not so dangerous if dropped, and tend to show distinct signs of wear before they fail, whereas an undetected fault in a glass vessel can cause a sudden failure.

Before autoclaving, fit a length of tubing and a suitable bung. The latter can be non-absorbent cottonwool, or a silicone rubber bung drilled to accept a length of tubing

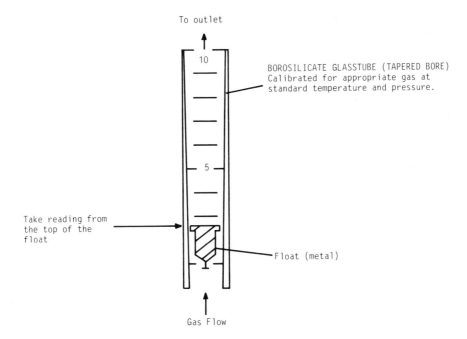

Figure 2. Operation of variable area flowmeter.

for medium pumping and a sterile venting system (see filters). Do not overfill the vessels. After autoclaving, allow vessel and contents to cool before removing them. Medium components likely to be affected by heat should be filter sterilized and added aseptically later.

5.9 Gas flow equipment

5.9.1 Measurement and regulation of gas flow

Most fermenters will be fitted with some means of controlling and measuring gas flow. At its simplest, this will consist of a needle valve giving very fine control of gas flow rate, provided supply pressure and batch pressure are steady, and a variable area flowmeter indicating the instantaneous rate of gas flow (*Figure 2*). Other means of measuring gas flow, such as orifice or vortex meters, are relatively rarely encountered on laboratory scale fermenters, though they have some advantages at larger scales.

These flowmeters are calibrated for particular gases (or liquids), air, oxygen, CO_2; the gas in use should be specified when the flowmeter is ordered. Where it is essential to ensure accurate and reproducible mass flow of gas, an automatic flow control valve should be fitted in addition to the variable flowmeter. These can be of two types; the first maintains constant mass flow of gas (or liquid) with varying upstream, but steady downstream, pressure provided a threshold level is exceeded; another type maintains constant mass flow at constant upstream pressure but variable downstream (batch) pressure. Since fluctuations in supply (upstream) pressure are very likely to occur, the first type of flow controller is advisable under most circumstances.

5.9.2 *Mass flow controllers*

These are very expensive, but are the most accurate means of ensuring a constant mass flow rate of gas regardless of pressure fluctuations upstream or downstream. They obviate the need for temperature/pressure compensations. They contain a heated filament and a temperature sensor; the amount of heat lost from the filament is proportional to the mass flow of gas over or past it, the relative reduction in temperature being measured by the sensor. Thermal mass flow controllers are so expensive that they should not be purchased unless they are essential for a particular application. They require a power supply, can be located at a distance from the controller, and may be linked to a computer for data logging control.

5.10 'O'-rings

Like tubing and clamps, O-rings are in constant use in a fermentation laboratory. They are used as the compressible material when a seal is made between glass and metal or between metal and metal. This allows the integrity of the seal to be maintained despite the glass or metal expanding and contracting at different rates. O-Rings are usually composed of nitrile or butyl rubber, sometimes of silicone. Since they are a standard industrial item, they are usually readily available from local sources (light engineering suppliers, etc.). It is financially well worthwhile establishing such a local source of supply, since fermenter companies charge very high rates for these essentially low-cost items. O-Rings fitted to the fermenter should be examined on a regular basis (at least monthly) for signs of damage or perishing. When fitting new O-rings, beware of over-compression which will cause the O-ring to flatten or splay out; such a squashed O-ring may not make an adequate seal. A smear of silicone grease may help prevent welding of the O-ring to the metal surfaces. Remember that items such as O-rings have a finite life span dependent upon how often they are autoclaved, and whether or not they are deformed by over-compression; they will thus have to be replaced at some point. If contamination is to be avoided, this replacement should be carried out as part of the regular maintenance programme.

6. SAFETY IN THE FERMENTATION LABORATORY

It is appropriate to start any consideration of laboratory techniques with a brief discussion of the safety aspects of fermentation. The brevity of the treatment does not imply that safety is less important than other topics. Safe working practices are usually also efficient ones; unsafe practices will lead not only to risks to health but also to disruption of experimental procedures. In order to avoid this, those in the laboratory must be aware of, and adhere to, a code of good laboratory practice; in particular, a regular programme of cleaning and equipment inspection should be instituted. The hazards associated with fermentation are similar in some respects to those found in other laboratories, but differ in the close proximity of the many potential dangers.

There are three major hazard areas.

(i) *Electrical*. Obviously there is a risk of shock or fire from the presence of large amounts of electrical equipment in the fermentation laboratory. The potential hazards are increased by the proximity of water, fermentation fluids and electrical equipment. Hazards are minimized by regular inspection and maintenance of electrical components, earths, plugs, cords, etc., by a competent person. In particular, residual current or earth leakage circuit breakers should be fitted to fermentation equipment wherever appropriate.

(ii) *Chemical*. This covers a wide range of potential or actual hazards, such as toxic medium constituents, e.g. phenylacetic acid, phenoxyacetic acid. In this case the operator must know how to safely store, handle, and dispose of the chemicals involved. A useful first contact is usually the supplier company. Another common hazard associated with fermentation is that posed by acid/base titrants and other corrosive agents. Wear gloves (disposable are best) and safety glasses at least when handling acid and alkali lines. Place acid and alkali reservoirs in a resistant container (e.g. polypropylene, polyethylene) rather than directly on the fermenter surfaces. This minimizes the risks associated with leakages of titrants into fermenter systems. Be especially careful if autoclaving titrant lines; again, place the reservoirs and lines in a resistant (and autoclavable) container and do not overfill reservoirs (not more than two thirds full). Ensure reservoirs are well vented, and inform those operating the autoclave that it contains corrosive materials. Take similar precautions with any toxic components to be autoclaved. It is not usual to autoclave titrants; instead the titrant reservoirs are usually partially filled with distilled water, autoclaved, then a concentrated solution of titrant is added to the sterilized water.

(iii) *Microbiological*. In most circumstances when dealing with microorganisms which pose a minimal hazard to laboratory staff or the environment, it is appropriate to insist on standard microbiological practices. These include:

(a)   restrict laboratory access;
(b)   decontaminate work surfaces daily and after any spillages;
(c)   autoclave all biological wastes;
(d)   use pipette aids: no mouth pipetting;
(e)   no eating, drinking, smoking, or application of cosmetics in the laboratory area. Do not store food, etc. in laboratory cabinets or refrigerators;
(f)   wash hands, preferably with a bactericidal soap, after handling viable agents and on leaving the laboratory;
(g)   carry out all transfers of cultures, etc. in such a manner as to reduce aerosol formation;
(h)   wear a laboratory coat or gown solely for microbial work.

For organisms in higher risk categories, more elaborate techniques are used. In the final analysis, it is the responsibility of the fermenter operator to find out the risk group of the microorganism and to take appropriate safety measures before commencing work.

Keep in mind that cultivation of microbes in (comparatively) large volumes increases potential hazards. Fermenters are usually vigorously aerated and agitated, which leads to extensive aerosol formation; they must, therefore, have a bacteria-proof exit filter. Similarly, fermenters operate with a positive headspace pressure which tends to force organisms outwards through any leaky points. To avoid this, check fermentation equipment, including tubing, regularly for damage, signs of wear and tear, obvious leaks, and repair or replace the item. Sterilize all fermenter contents before discharge and clean the fermenter and surroundings thoroughly between runs.

Books on the subject of safety abound, but a particularly useful one covering aspects of good laboratory practice is *Laboratory Biosafety Manual*, WHO, Geneva, 1983 which covers most points well and succinctly.

In addition to the hazards indicated above, there is always a risk involved in the use of live steam and compressed gases. The golden rule in safety is to find out how to do it safely *before* you start work. Every new laboratory worker should, therefore, be instructed in laboratory safety methods before being allowed to use any equipment.

7. FURTHER READING

1.   Miller,B.H. (ed.) (1986) *Laboratory Safety: Principles and Practices*. American Society for Microbiology, Washington, DC.
2.   Collins,C.H. (1988) *Laboratory-acquired Infectious Infections* (2nd edn.). Butterworth, London.
3.   Fuscaldo,A.A., Erlick,B.J. and Hindman,B. (1980) *Laboratory Safety: Theory and Practice*. Academic Press, New York.

CHAPTER 2

Laboratory fermenters

THOMAS S.IRVINE

1. INTRODUCTION

The vast majority of industrial fermentation companies have at least a research and development section dedicated to laboratory fermentation. These laboratories are used for pure research for new products, screening for new fermentation cultures, developing new raw materials for fermentation, process trouble-shooting, and scale-up. Examples are to be found in the pharmaceutical industry where new antibiotics are developed, in the brewing industry, or in the food industry where new fermented foods such as protein substitutes are developed.

In these research and development sections the type of laboratory fermentation may vary from a small 100 ml shake flask to a large 100 l stainless steel fermenter; in a large development unit there will be several sizes of mini-fermenter, starting from one to 5 litres, scaling up to 20−50 litres and, finally, pilot plant where the scale up could be 100−1000 litres. The number of larger mini-fermenters is usually governed by costs and manpower availability. It should be noted that the definition of what constitutes a bench-top, laboratory, or pilot-plant fermenter varies from user to user. In simple terms, bench-top fermenters have working volumes up to 3 litres, the vessel is almost invariably glass, and they are usually not sterilizable *in situ*; laboratory fermenters are usually in the range 5−50 litres working volume, are free standing, and normally sterilizable *in situ*. The vessel can be of stainless steel or glass, or a combination of both.

In recent years microbiologists and chemical engineers have developed several sophisticated mini-fermenters capable of specializing in different types of fermentation. There are numerous fermenter types. These include stirred tank reactors (STR), air-lift fermenters, tower fermenters, fluidized bed reactors and rotating disc fermenters. Each of these configurations of fermenter may offer advantages in certain applications. This is discussed more fully later in the chapter.

2. FERMENTER AND BIOREACTOR TYPES

Historically, the main tool for a fermentation laboratory was the shake flask or flat-bed bottle. In fact, after the Second World War when the penicillin fermentation was being developed, hosptial bedpans were commandeered to allow the development scientists to continue with their studies.

Science has made great strides since then, but the shake flask and flat-bed bottle still have an important role to play in modern fermentation laboratories. The next stage was the introduction of glass vessels with a stirrer, and this was followed more recently

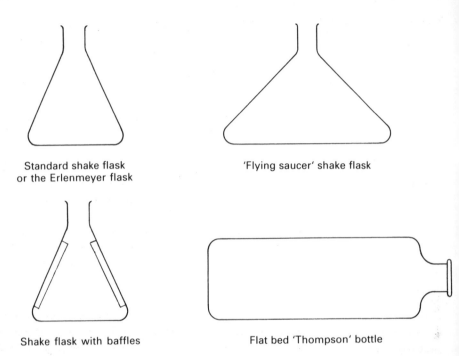

Standard shake flask
or the Erlenmeyer flask

'Flying saucer' shake flask

Shake flask with baffles

Flat bed 'Thompson' bottle

Figure 1. Shake flasks for fermentation. (**a**) Standard shake flask (Erlenmeyer flask); (**b**) 'Flying saucer' shake flask; (**c**) shake flask with baffles; (**d**) Flat-bed 'Thompson' bottle.

by stainless steel vessels in various sizes and forms, an example of the latter being the stainless steel mini-fermenters (stirred tanks) which, in essence, are scaled-down versions of 100 m^3 production-scale fermenters.

2.1 Shake flasks and bottles

These pieces of glassware can vary in size and form, and in some instances have been designed and developed for specialist application. *Figure 1* gives some idea of the shape and type of flasks used in the fermentation laboratories.

Shake flasks find a variety of uses in the fermentation laboratory, including initial strain screening, multivariable testing, and inoculum growth. The subject of inoculum preparation is dealt with in Chapter 3. Normally in preparing the inoculum for a fermenter at least one stage will involve cultivation of the organism in shake flask. Shake flasks have a number of significant disadvantages relative to cultivation in a fermenter, including lower oxygen transfer rates, less closely controlled environmental conditions, and the difficulty of withdrawing samples aseptically.

Oxygen has a low solubility (9 p.p.m. O_2 saturation level at 25°C) which means there is a resistance to transfer between the air bubbles and liquid. The rate of transfer of oxygen into a liquid can be described by the following equation:

$$N_a = k_L a \, (C_g{}^* - C)$$

where N_a is the oxygen uptake rate (mmol litre^{-1}h^{-1}), k_L is the mass transfer

Table 1. Typical values of $k_L a$ in various systems

Fermenter	$k_L a$ value (h^{-1})
Test-tube	20
Flat-bed bottle	50
Shake flask	500
Shake flask (with baffles)	1200
Mini-fermenter	3000 – 4000

coefficient (litre h^{-1}), a is the interfacial area for mass transfer (litre^{-1}), C_g is the saturation level for O_2 in the liquid (mmol litre^{-1}), and C is the actual dissolved level (mmol litre^{-1}). $k_L a$, the mass transfer coefficient, has to be calculated for each fermentation apparatus.

Table 1 gives some indication of the values of $k_L a$, and how important it is.

The main constraint in shake flasks is oxygen transfer rate (OTR), which is governed by $k_L a$. Various methods can be adopted to increase shake flask efficiency.

2.1.1 *Shaker tables*

Shaker tables were designed to assist with oxygen transfer. These tables are designed to run for long periods of time and be free from vibration. The tables are driven by a motor, and normally a rotary shaking action or reciprocating shaking action is produced.

These shakers have to be robust and reliable with no vibration and silent running conditions. One can have a more sophisticated shaker by having an incubator shaking cabinet for shake-flask fermentation in a precisely defined environment. These cabinets can control the temperature, illumination, gaseous levels, and humidity.

Increasing the speed of a shaker can increase the oxygen transfer rate of a particular flask, therefore the optimum speed for that flask and culture has to be found by trial and error.

2.1.2 *Shake flask volume*

The lower the volume of medium in a shake flask, the better will be the OTR. The volume will be dependent on the medium and type of culture. The minimum volume that can be practically obtained (e.g. 50 ml in a 250 ml shake flask) should give the best OTR and hence the best results. This will also be dependent on sample volume. Very low volumes can only be used for short-term fermentations, otherwise the medium will evaporate and the nutrients would become too concentrated for the culture to perform satisfactorily. The smaller the volume, the thinner the film on the flask and the greater the area in the $k_L a$.

2.1.3 *Fermentation glassware*

The standard 250 ml Erlenmeyer flask is cheap and simple; most of the shaker tables are designed to use these flasks, although there are tables which can be adapted to allow other shapes or bigger flasks.

Baffles have been used in shake flasks to assist in the OTR, as well as preventing vortex formation, but, they are only really suitable for low-volume, short-term fermentations beause of splashing which leads to the cotton-wool plug becoming damp preventing free flow of oxygen.

Different plugs can be made of cotton-wool, glass wool, polyurethane foam, gauze or synthetic fibrous material. (An aluminium foil cap can sometimes be used in conjunction with these plugs.) The plug has to prevent airborne microorganisms from getting into the medium while at the same time allowing free flow of air into the flask, and for this reason it must not be allowed to become wet.

Other shapes of glassware are used, such as the 'flying saucer' shake flask or the 'Thompson' flat-bed bottle, but these are very expensive compared to the standard 250 ml shake flask and tend to be used only in specialist applications, for example, gluconic acid or antibiotic fermentations.

2.2 **Stirred tanks**

Basically, the stirred tank fermenter consists of a cylindrical tube with a top-driven or bottom-driven agitator. The stirred tank with a top-drive assembly is the most commonly used fermenter because of its ease of operation, neat design, reliability, and robustness. For smaller mini-fermenters (bench-top), borosilicate glass is used as the cylindrical tank and a top plate of stainless steel clamped on. A motor is fixed above the top plate and is attached to the shaft. The motor can be uncoupled and the mini-fermenter is readily sterilized in an autoclave. The vessel, medium and probes are usually sterilized together, minimizing the number of aseptic operations required.

For safety reasons, glass vessels are normally sterilized in an autoclave, although some can be sterilized *in situ* if the necessary safety precautions are taken. The glass vessel can be protected by a removable stainless steel mesh, or jacket.

These glass vessels can vary in size from one litre to 30 litre capacity. The vessel itself will have a specific impeller design, baffles, an air sparger, and sample port(s). *Figure 2* shows the basic configuration.

The stainless steel minifermenter is the 'Rolls Royce' of laboratory fermenters. It is a hollow steel cylinder with either top or bottom drive and can be cleaned and sterilized *in situ*. These stirred tanks can vary in volume from one litre to 100 litres capacity. Obviously they are more expensive than the glass vessels but they are more robust, reliable, and designed to last a lifetime.

Often large samples or regular samples have to be taken for analysis during the fermentation. These volumes must be taken into consideration when choosing a fermenter. It would not be practicable or worthwhile to purchase a one-litre fermenter if the assays to be carried out daily required a 100 ml sample.

The size and shape of these vessels will be studied later. *Figure 3* shows an outline diagram of a basic stainless steel vessel. They are designed to give variable speed agitation; thus, if required for certain aerobic mycelial fermentations such as *Penicillium chrysogenum*, a great deal of power can be put in to promote mixing and oxygen transfer. Lower power and less agitation could be used for a bacterial fermentation.

These stainless steel laboratory fermenters can be trolley mounted or skid mounted, depending on size and situation. The vessel itself should be made from high-grade stainless steel with the internal surface polished smooth to minimize microbial adhesion to the vessel sides. There is controversy as to whether or not mirror finish is essential

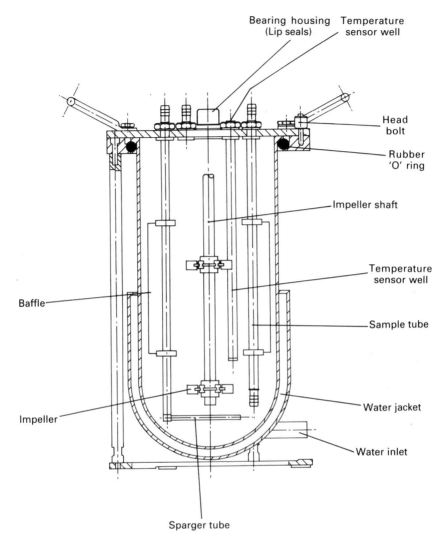

Bearing housing Temperature
(Lip seals) sensor well

Head bolt

Rubber 'O' ring

Impeller shaft

Temperature sensor well

Baffle

Sample tube

Water jacket

Impeller

Water inlet

Sparger tube

Figure 2. (a) Typical glass fermentation vessel (Courtesy of Life Science Laboratories Ltd).

on the inside of vessel walls. Welded joints must be to a very high standard and free from any pin-holes.

An external jacket is welded onto the vessel to provide accurate temperature control; the use of a jacketed immersion heater and/or a cooling finger is less attractive.

2.2.1 *Automation and performance*

Bench-top fermenters are usually cheaper to purchase than the trolley-mounted or skid-mounted fermenters. This is partly due to the fact that their instrumentation is often not as sophisticated as laboratory or research fermenters. The latter have a sophisticated

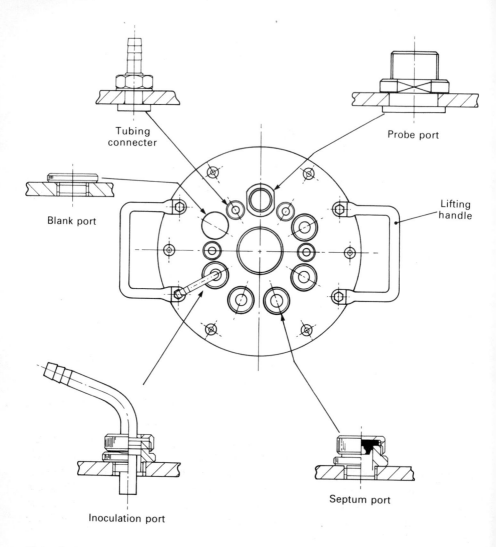

Figure 2. (**b**) A typical vessel headplate (Courtesy of Life Science Laboratories Ltd.).

instrument control package for pH, temperature, and agitation, and this obviously costs more.

Although a mini-fermenter may be reliable in terms of sterility, etc., if it does not have a reliable instrumentation package, the fermenter will not achieve its full potential. Chapter 7 deals with instrumentation and control in greater detail.

2.2.2 *Agitation and aeration*

Most stainless steel fermenters are designed with a bottom drive unit which is belt driven. This has several advantages:

(i)   It allows easy access to the top of the vessel and all the moving parts can be isolated and encased underneath the vessel, thus making it a safer piece of apparatus;

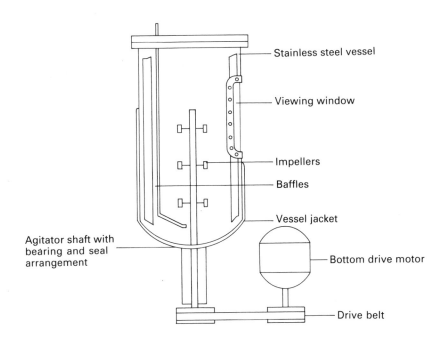

Figure 3. Stainless steel fermenter.

(ii)   With the motor belt driving the agitator, any spillages which occur will not fall onto the motor because it is not sited directly below the agitator shaft.

The agitation shaft should have a double mechanical seal which ensures that the medium does not leak out at the shaft housing. The agitation shaft will normally have two or three impellers, each with four or six blades depending upon mixing requirements.

The air supply is filtered by one of the methods discussed in Chapter 1. These can filter particles down to 0.2 μm size, thus providing sterile air. The sterile air is then fed into the bottom of the fermenter dispersed by a sparger and thoroughly mixed into the medium by the agitation system.

Vent gases can be filtered by the same means, but one has to be aware of the risks of blockage due to carry-over of medium or foam-out. Such risks can be minimized by the use of a foam control system; either a mechanical system (offered as an option by some fermenter manufacturers) or, more routinely, a system involving addition of some foam-suppressing chemical (e.g. a silicone-based compound or polypropylene glycol). An efficient condenser fitted to the gas outlet will also reduce the likelihood of exit filter blockage. Exit or vent filters (0.2 μm or 0.45 μm pore size) should be fitted to all fermenters. The use of surge tanks is also advisable.

2.2.3 *Ancillary equipment*

Ports for pH, dissolved oxygen, and temperature sensors, can either be mounted on the sides of the vessel or on the top of the vessel. Feed ports are situated on the top plate of the fermenters along with an inoculation port. On larger vessels (>20 litres)

23

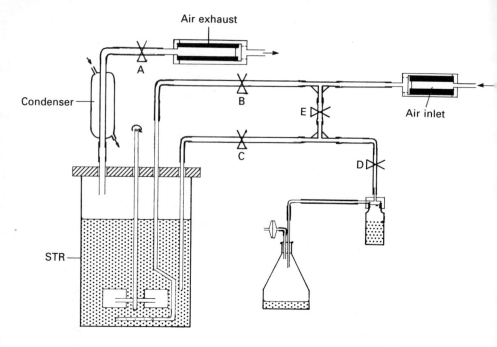

Figure 4. Illustration of a technique for taking small-volume samples from a laboratory STR.

an illumination lamp is usually fitted on the top plate to allow visual examination of the culture broth.

Since the vessel is pressurized during sterilization, there is a pressure gauge mounted on top of the vessel and a manual vent valve can be adjusted to control the pressure setting required. A relief valve is fitted to the vessel top to prevent overpressurization of the vessel. A sample pipe will also be fitted on the top or side of the vessel.

The frequency of sampling is governed by the design of the experiment and the sample size in relation to the culture volume. Overall, not more than 10% (v/v) of the original volume should be removed, otherwise physical conditions such as oxygen transfer rates will be affected. For example, in a batch culture time-course experiment hourly samples may have to be taken over a 24 hour period. If samples are taken in universal bottles (approx 30 ml per sample), then the overall sample volume will be around 720 ml: an amount unacceptable for culture volumes of less than 10 litres.

2.2.4 *Sampling devices*

Different fermenter designs have different sampling devices. The larger *in situ* steam-sterilizable fermenters usually have a built-in lever-type sample port which can be readily sterilized before and after every sample (ports should be steamed for at least 5 min). Bench-top fermenters, especially those with no facilities for bottom access, usually requires more elaborate sampling device. One such device is illustrated in *Figure 4*. This sampling device is linked up to a sterile air stream so that the tubing can be flushed free of residual cells before and after sampling. Under normal running conditions, clips A and B are open and clips C, D, and E, are closed.

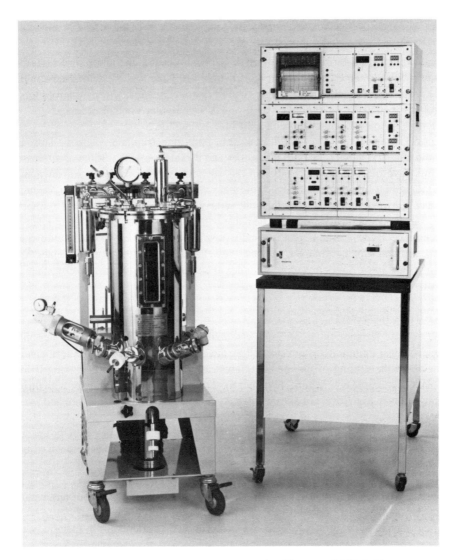

Figure 5. A 20 L stainless steel fermenter and instrumentation (Courtesy of LSL Ltd.).

The method of sampling is as follows:

(i)   Open E and C. Close B and allow the sample tube to flush free of any residual cells from the previous sample;

(ii) Close A and E then carefully open D and allow the culture to flow into the universal bottle displacing any trapped air into the waste receptable. Alternatively, the cap of the universal can be slightly loosened to bleed off excess air. Close D;

(iii) Carefully open A and slowly reduce the pressure in the fermenter;

(iv) Open E and allow air to blow through the sample tube;

(v)   Open B then close E and C;

(vi)   Using aseptic techniques, remove the sample bottle from the assembly and replace with a fresh sterile universal bottle.

This type of mini-fermenter has been developed to enable the laboratory researcher to work with fermenter volumes previously only available to those with access to pilot scale processes. The automated systems can be operated manually or by microprocessor cotrollers mounted in a synoptic panel or via a computer control module.

Figure 5 shows a photograph of a 20 litre fermenter.

The STR fermenter is very widely used in both industrial and academic labs. It is still the industry standard at both laboratory and production scale. Some of the reasons for this have been discussed earlier in this section, but one of the major advantages of STRs is their inherent flexibility. Almost any type of microorganism, plant, or animal cell can be grown in what is essentially the same vessel. Minor modifications to impeller type, stirrer speed, airflow rate, can be made which permit cultivation of even relatively fastidious cell types. Similarly, it is usually possible to employ any mode of cultivation, batch, fed-batch or continuous, in these vessels with few alterations (Chapter 5).

However, since STRs are more complex than air-lift and tower reactors, they are usually more costly.

2.3 Air-lift fermenters

In all aerobic fermentations, air is an essential requirement. Air-lift fermenters have no mechanical agitation system but utilize the air circulating within the fermenter to bring about the mixing of the medium. This rather gentler system of mixing is ideal for plant and animal cell cultures (Chapter 9). The standard mini-fermenter with turbine mixing leads to high levels of shearing which could cause cell lysis in plant and animal cells.

The air-lift fermenter is designed in such a way that aeration provides both the agitation of the broth and dispersal of the oxygen into the broth. Power draw is very much reduced and the main power source is the air compressor.

Antibiotic fermentations using fungal cultures require large quantities of air-enriched medium and, therefore, air-lift fermenters are not as efficient for this type of fermentation since they have poorly defined mixing patterns compared to an STR.

The principle of the air-lift fermenter is based upon the differences in specific weight between the air-enriched volume and low-air volume. As the fermenter is aerated, the lower density broth (air-enriched medium) creates an upward thrust which results in the circulation of the broth. The type of circulation is dependent upon the vessel configuration, as *Figure 6* outlines.

The basic design for a laboratory scale air-lift fermenter is an outer glass hollow tube with an inner stainless steel tube.

A variation of the air-lift fermenter principle is the tubular loop fermenter. This type of fermenter can be used to increase the volume of the fermentation while maintaining the residence time. They can be run in series, but *Figure 7* shows a single tubular loop fermenter. Design and performance of these reactors is presented in reference (1).

2.4 Tower fermenters

Tower fermenters were developed for continuous yeast fermentation processes.

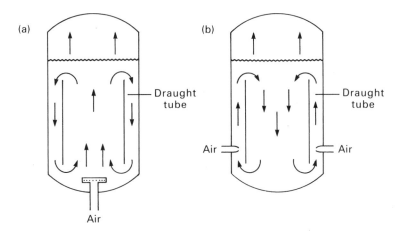

Figure 6. Two types of air-lift fermenters: (**a**) Air is fed at the base of the fermenter and the arrows depict the flow of medium/air. (**b**) Air is fed in at the sides of the vessel and the arrows depict the flow of medium/air.

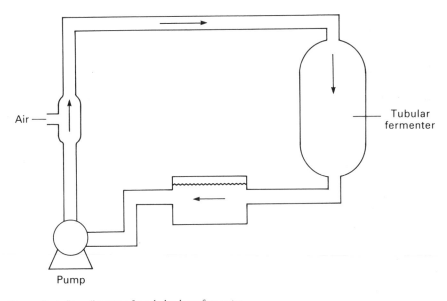

Figure 7. A flow diagram of a tubular loop fermenter.

Continuous brewing of beer or lager can be carried out successfully in a tower fermenter. The design of these fermenters is fairly simple, and they are less expensive than a conventional STR. 30−50 litre capacity tower fermenters can be used in the laboratory for research and development. *Figure 8* outlines a typical tower fermenter for a yeast fermentation.

This design concept was used in continuous production of single-cell protein. Airlift and tower fermenters, due to their lack of complex mechanical agitation systems, are easier to construct at laboratory scale than STRs.

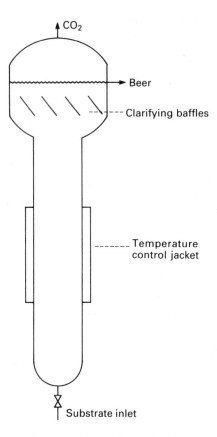

Figure 8. A typical tower fermenter for a yeast fermentation. pH and temperature probes are situated at the top and side of the vessel. The yeast particles (flocs) are fluidized by the upward movement of the broth and the flocs are retained by the clarifying baffles, disrupted and descend to the bottom of the tower.

2.5 Bioreactors utilizing immobilized biomass

Many other types of bioreactor have been developed and used in the laboratory. However, some of these designs are just not practicable or economic to scale up to production scale. On the other hand, old methods such as trickling filters have been developed further, and the fluidized bed has been incorporated into a bioreactor.

The immobilized biomass consists of microorganisms which adhere to larger particles such as rocks, glass beads, or plastic beads. The quantity of biomass is dependent upon the inert particle surface area, sloughing effect, aeration, and recycle efficiency.

2.5.1 *Fixed bed reactors*

The trickling filter has been used in waste water treatment for almost a hundred years now. Inert pieces of stone, slag, or brick are used for the microbial biomass to adhere to. This type of fermentation is a heterogeneous reaction, whereas the stirred tank fermentation (ideally, at least), is a homogeneous reaction.

Several systems have been developed in the laboratory with a tubular packed bed

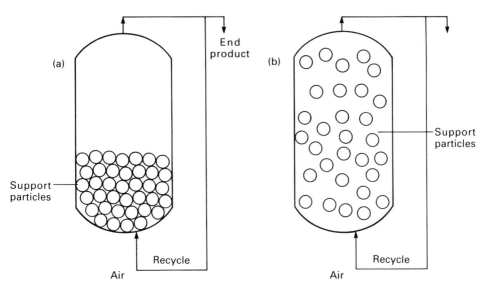

Figure 9. Packed bed and fluidized bed bioreactors. (**a**) the packed bed reactor has a restricted flow rate compared to (**b**) the fluidized bed reactor.

reactor, one of which is shown in *Figure 9*. The main problem associated with these reactors is getting the fixed bed fully aerated. If air is restricted then anaerobic microorganisms take over.

2.5.2 *Fluidized beds*

The use of inert dense particles has been further developed in fluidized beds. These reactors are hollow chambers in which dense particles containing a microbial film or microbial mass are mixed or recycled. A typical example is shown in *Figure 9b*. One of the oldest fermentations known to man, vinegar fermentation, utilizes this principle. The alcoholic broth is recycled through a hollow tube reactor with dried beechwood shavings. The bacterial film of *Acetobacter* species grows on the surface of the wood shavings.

2.5.3 *Rotating disc fermenter*

Waste water treatment has been and still is effected by the use of rotating discs. The microorganisms adhere to the discs and these rotate slowly through the waste water, so the microbial film is exposed both to a nutrient solution and to air.

Blain *et al.* (2) have designed a small-scale disc fermenter for filamentous fungi fermentation (*Figure 10*).

Further development work is being undertaken to explore the potential of disc fermenters in industrial fermentation applications.

2.6 **Design and construction principles of a stirred tank bioreactor**

Microorganisms can grow in any shape of vessel. Each microorganism has its own

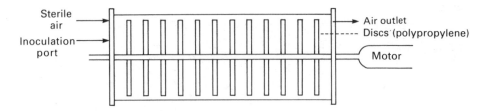

Figure 10. A basic laboratory disc-fermenter.

individual requirements. Whether it be environment and/or nutrient supply, the designer must take certain constraints into consideration:

(i) the culture must be protected from other microbial contamination and, therefore, the vessel must be able to withstand sterilization temperatures;

(ii) all feed supply pipes must remain sterile either by autoclaving or by *in situ* steam sterilization;

(iii) valves must be used which are designed to sterility standards; and

(iv) all connections should be welded rather than flange bolted together.

In small-scale set-ups this is not practicable and silicone tubing is used instead of stainless steel pipe—connections on smaller vessels are, therefore, made with tubing connectors (see Chapter 1).

Good-quality stainless steel should be used for the vessel construction, 316 S.S. grade steel or equivalent being the most suitable. All connections should be precision stainless steel welding because these vessels have to withstand high temperatures and pressures, different pHs, different media, and rigorous cleaning schedules.

Cell morphology must be taken into consideration when designing a fermenter. The type of agitation and aeration can have quite an effect on morphology. An example of different morphology can be seen in some filamentous fungi. In submerged culture, these cultures can produce small compact growth centres or pellets (*Figure 11*). These pellets will disperse and form a filamentous mat but this will be dependent upon agitation and shearing action.

This leads onto another topic the vessel designer has to consider; the rheological properties of the fermentation broth. The viscosity can be either Newtonian or non-Newtonian. Most bacteria and yeasts in submerged cultures exhibit Newtonian characteristics, so increasing the shear rate does not have an effect on the viscosity. However, cultures of filamentous fungi do exhibit non-Newtonian characteristics and shear rate does have an effect on apparent viscosity as shown in *Figure 12*; increasing the stirrer speed can reduce the apparent viscosity of the culture.

The rheological properties of filamentous cultures depend on the cell concentration, hyphal length and branching, flexibility of mycelium, and cell morphology. Most filamentous fermentation broths are referred to as pseudoplastic fluids but experimental work has shown that some of these broths are, in fact, Casson fluids or Bingham plastics. *Figure 13* outlines the various rheological properties demonstrated by typical fermentation fluids.

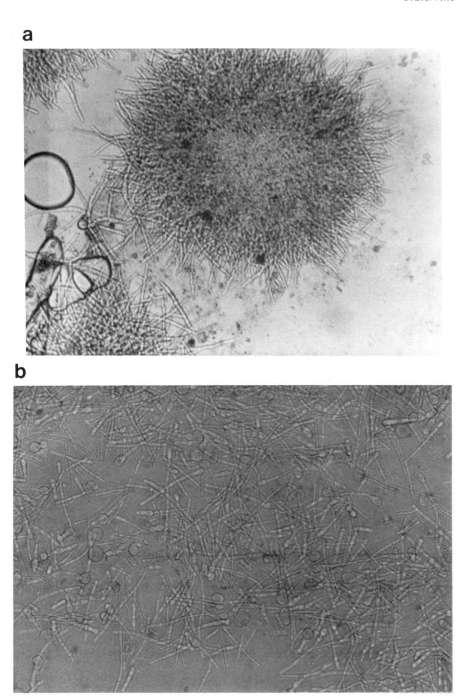

Figure 11. (a) Pellet. (b) Filamentous mycelium.

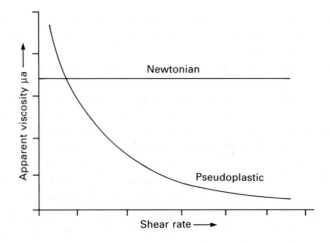

Figure 12. Shear rate dependence of apparent viscosity for Newtonian and pseudoplastic broths.

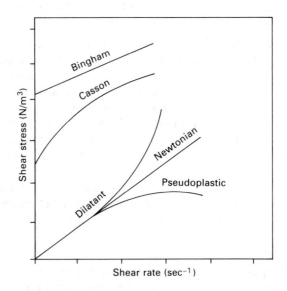

Figure 13. Fluid rheology characteristics (Solomons, 1969).

3. MIXING AND AERATION

3.1 **Power requirements**

In a stirred tank bioreactor an electric motor is required to drive the agitation shaft. The motor must generate enough power to ensure that the fermentation medium remains a homogeneous mix. For a small reactor of (20 litres), a 1 kW motor is adequate.

The motor can be connected to the fermenter shaft in a number of ways:

(i)   *Direct coupling*. This is the standard method for a top-drive assembly.

(ii) *Belt-driven coupling.* This is normally used for bottom-drive fermenters because the motor can be sited away from the fermenter and any spillage is not going to get into the motor.

(iii) *Flexible coupling.* Usually made of vulcanized rubber, this compensates for motor and shaft play.

(iv) *Magnetic coupling.* Multipole ceramic magnets are used to drive the agitation. The main advantage is that it eliminates a possible source of contamination through the shaft housing assembly. Whether or not this type of coupling can be used depends upon the rheological properties of the fermentation fluid. The use of such a system obviously limits the future applications of the fermenter.

(v) *Gear coupling.* Gear assemblies can be used normally in conjunction with flexible couplings to enable one motor to drive three or six fermenters.

The impeller drive shaft enters the fermenter vessel via either the top or bottom plate. The shaft must be free to turn (low friction) and the assembly must not allow the medium to escape or microbial contamination to enter.

The shaft is allowed to turn freely by the use of roller ball bearings and, in some instances, a bottom or steady bearing can also be installed to prevent any 'whiplash' action. There are a number of different seal designs used to prevent contamination:

(i) *Mechanical seal.* Usually these seals are made of a ceramic material or carbon graphite, and if lubricated properly they can last for a long time. They can be lubricated with glycerol or steam; the latter also acts as a contamination guard. Another design utilizes sterile air which is then passed over the mechanical seal.

(ii) *Rubber seals.* These are known as lip seals, and can be made of butyl or nitrile rubber and may be used in conjunction with a mechanical seal as a safeguard. The material used must be able to withstand repeated sterilization and wear and tear of the shaft. A regular programme for inspecting and replacing these seals should be adopted.

(iii) *Gland packing.* This is an asbestos type of material, usually used in larger fermenter drive shafts. It is would round the shaft and then compressed by an outer bracket to complete the seal.

3.2 Impeller designs

Various forms of impeller can be used in stirred tank bioreactors.

Dissolved gases such as oxygen must be mixed properly, and the evolved carbon dioxide must be removed by the mixing effect. Carbon dioxide build-up can lead to product inhibition in many aerobic fermentations. *Figure 14* outlines the main designs used for impellers. There are numerous variations from these designs, e.g. the disc turbine may have four blades instead of six if less power is desired, and the blades themselves may be set at an angle. The marine propeller is often used for plant and animal cell culture because it does not create the high shearing effect of turbine impellers (Chapter 9).

The Rushton blade disc turbine has traditionally been the most popular impeller. A lot of power is required to drive it, but it is simple in design and robust in construction. It can easily be replaced or altered for experimental purposes.

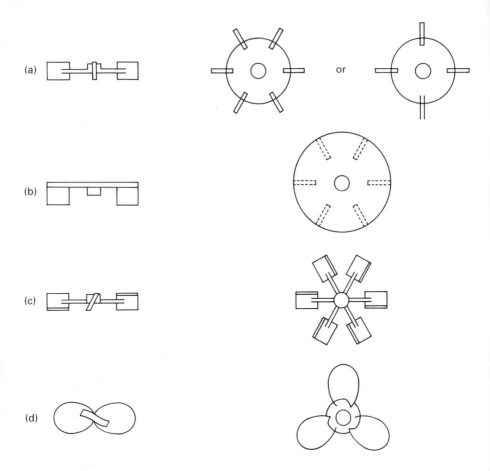

Figure 14. Types of impellers: (**a**) Disc or Rushton turbine with six or four blades. (**b**) Varied disc; (**c**) Open turbine, variable pitch; (**d**) Marine propeller (Solomons, 1969).

3.3 **Baffles and aeration**

Baffles are normally fixed in an offset position on the vessel sides. Usually four baffles per vessel are used to prevent a vortex formation. If the baffles are not fixed in an offset position (*Figure 15*), then stagnant zones are formed in the fermenter, causing a build-up of microbial biomass.

The baffles and impeller configuration produce either an axial or radial flow pattern, depending upon type of impellers used, as shown in *Figure 16*. The most effective mixing is achieved by using a turbine impeller with baffles.

An internal cooling coil can have the same effect as baffles and preventing vortexing.

For aerobic fermentations, the air is normally supplied by compressor(s) and the air is fed into the vessel under pressure. The air is fed via a sparge pipe directly below the impeller zone. The types of sparger can vary considerably and *Figure 17* outlines the most commonly used spargers. The most efficient sparger is a sintered stainless

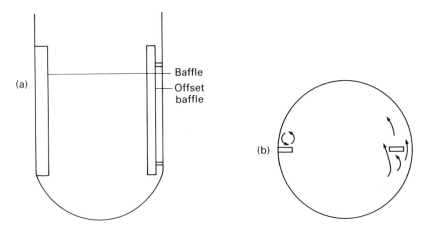

Figure 15. (**a**) Two methods of attaching a baffle to the fermenter. The offset baffle is usually preferred. (**b**) Baffle attached to the vessel wall leads to stagnant eddy currents described by arrows in the diagram.

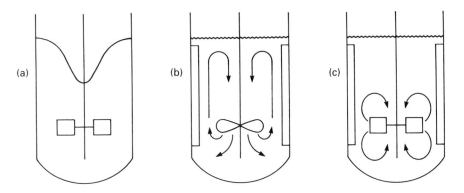

Figure 16. (**a**) Turbine impeller; large vortex, no baffles. (**b**) Marine impeller; axial flow with baffles. (**c**) Turbine impeller; radial flow with baffles.

steel disc, but this can be prone to blockage. The most useful one is normally the ring sparger or, for mycelial organisms, a single tubular sparger.

It is essential to get the air to dissolve into the liquid as efficiently as possible. The ring sparger wih perforated holes can distribute the air very efficiently but the disadvantage is that the holes are prone to blockage.

4. VESSEL CONFIGURATION

Fermenter capacity is defined as follows:

Total fermenter volume = Working volume + Head space volume

The working volume of a fermenter is that fraction of the total volume filled with culture (medium, microoganism, and gases present in the liquid: the gas hold-up). Since

35

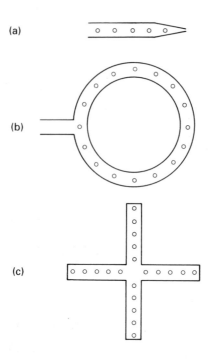

Figure 17. (**a**) Dip pipe: less prone to blockage. (**b**) Ring sparge: this can cover a greater area in the fermenter and ensure a better dispersal of air. (**c**) Spider sparge: this design can rotate inside the vessel ensuring a greater dispersal of air.

space has to be allowed at the top of the tank for gas disengagement and foaming, it is usual to find working volumes for STRs around 80% of total volumes.

Original stirred tanks had a height/diameter ratio (H/D) of between 1:1 and 2.5:1. Modern designs of stirred fermenters have H/D between 2:1 and 3.5:1, with several impellers mounted on the drive shaft. For a vessel with H/D less than 2:1, one impeller is satisfactory.

The preferred shape of the stainless steel vessel is normally cylindrical with a hemispherical or 'dished' end at the bottom and a flat top plate as shown in *Figure 18*. The dished end has several advantages: for instance, all liquid can be drained from the vessel. This design, as shown in *Figure 18*, eliminates any 'dead space' or stagnant areas which are undesirable.

5. BUYING A FERMENTER

There are many good manufacturers of mini-fermenters throughout the world. Some deal in glass vessels, others only in stainless steel, and some in both. There are numerous colourful brochures for the intending purchaser to browse through.

Purchasing a mini-fermenter involves a large capital expenditure, anything from a few thousand pounds for a small bench-top fermenter to tens of thousands of pounds for some of the larger STRs. The equipment must be from a reliable manufacturer, because it has to have a working life of anything from 10 to 20 years.

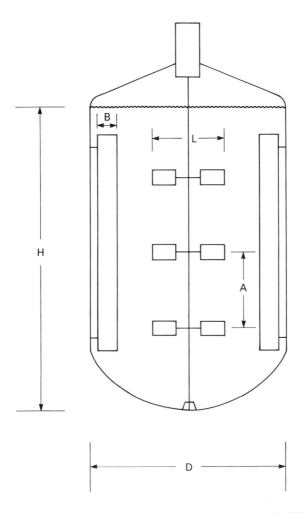

Figure 18. H/D = 1.1 to 3.5:1, L/D = 0.25:1 to 0.5:1, A/L = 1:1 to 1:1.2, B/D = 0.1:1 to 0.2:1.

The list below gives a general guide to some of the points worthy of consideration when purchasing a laboratory fermenter:

(i)   The fermentation process must be compatible with the manufacturers' design; for example, high power and shear rate for a filamentous fungal culture. Most manufacturers will supply a custom-built model, but extras will add to the cost.

(ii) It is advisable to see the type of fermenter you are buying either at a trade exhibition by arranging to see one running which has been purchased by another customer. Better still, the manufacturer might arrange a demonstration for you and let you assess the fermenter personally.

(iii) The manufacturer will supply a list of services which are required (Chapter 1). Make sure these are available (e.g. steam, air, water, electrical supply). Check that the water supply pressure is adequate and that it is cool enough to be used as a coolant in both summer and winter.

(iv) Instrumentation and control is discussed in Chapter 7; ensure that the control instruments are compatible with your computer system.

(v) Always get a quotation in writing, and make sure it is in local currency because of currency fluctuations. Such a quotation should specify exactly what you are getting for your money, what kind of guarantee period is being offered, and the delivery time.

(vi) If you enter into a service contract, find out exactly what you get per annum; in particular, find out if the company has a good back up-team and if they have an adequate spare parts list. Also find out how long it takes to deliver the spare parts. A reliable company will have a good after-sales service; be wary of companies whose after-sales service is likely to be poor.

(vii) Many long-term projects nowadays are undertaken jointly by university and industry; therefore, when the need to purchase equipment arises the two parties can halve their costs.

(viii) Many manufacturers offer lower priced 'options' with reduced amounts of instrumentation (e.g. no D.O. controller or meter). These items may not always be essential at first, and in most cases can be acquired as options later. As mentioned previously it is probably not advisable to buy low-quality instrumentation packages since this is likely to render the vessel inoperative more frequently than is desirable.

Finally, it must be said that buying a laboratory fermenter represents a considerable financial investment, but buying the best does not necessarily mean paying the most. The warning that 'all that glistens is not gold' holds good here too.

6. REFERENCES

1. Katinger,H.W.D. (1977) In *Biotechnology and Fungal Differentiation*, J.Meyrath and J.D.Bu'Lock (eds), Academic Press, London, p. 137.
2. Blain,J.A., Anderson,J.C., Todd,J.R. and Divers,M. (1979) *Biotechnol. Letts*, **1**, 269.

7. FURTHER READING

Atkinson,B. (1974) *The Principal Types of Fermenter. Biochemical Reactors.* Pion, London.
Kristiansen,B. (1978) *Chem. Ind.*, 21 October, 787.
Solomons,G.L. (1969) *Materials and Methods in Fermentation.* Academic Press, London.

CHAPTER 3

Strain preservation, inoculum preparation and development

CAROL PARTON and PETER WILLIS

1. INTRODUCTION

Conventional modern microbiology depends on the availability of pure, stable cultures. Unless microorganisms receive expert handling they are vulnerable to contamination, spontaneous mutation, deterioration and death. Most industrial production strains are irreplaceable; many have been developed by empirical methods and are often inherently unstable. In cases where cultures can be re-isolated or replaced by service collections the associated costs in time and effort are undesirable, so the maintenance of adequate viable stocks is essential. The financial implications of operating industrial processes with strains that have deteriorated are obvious. Decrease in secondary metabolite titre, associated with deterioration of the primary culture, is well recognized. For example, serial sub-culture over 7 – 10 steps of inoculum cultures of *Penicillium chrysogenum* and *Streptomyces niveus* results in a total loss of production of penicillin and novobiocin respectively. Similar problems can result from the mishandling of organisms at later stages in a fermentation process. Losses can be avoided or minimized by employing effective systems for culture preservation, inoculum preparation and fermentation scale-up.

2. STRAIN PRESERVATION; CULTURE COLLECTION MANAGEMENT

A wide variety of techniques is available for the preservation of microorganisms. All methods have their advantages and disadvantages, and no single method for the successful preservation of all microorganisms exists.

At least one of the methods to be discussed later should give satisfactory results in most cases. The choice of method should be determined by relating the features of the method to the requirements of the microorganism and the needs of the user. Factors which should be taken into account when selecting the method of preservation include the degree of viability required on preservation, the consequence of population and genetic changes on the organisms, the number of samples requiring preservation, the frequency of culture withdrawal and the unit cost of the method chosen. The nature of the culture and the type of equipment available in the laboratory may also determine the preservation method of choice.

Desiccation of spores and mycelium on to sterile garden loam is a convenient method for many fungal and actinomycete species. The need for specialized equipment is minimal and once the culture has survived desiccation it will continue to survive for a number of years if the soil remains dry. Sporing species preserved in this way often survive longer than non-sporing species.

Other methods of preservation by desiccation include desiccation on sand, silica gel, paper strips and gelatin discs. These methods have proved useful for some groups of yeasts, fungi and bacteria. Capital equipment costs are small and none of the methods is unduly labour-intensive.

Many fungi, actinomycete and bacterial species may be stored frozen. The cells are suspended in glycerol for protection and then held at a low temperature. The cells can be held in a freezer between $-50°C$ and $-70°C$, or in the vapour phase of liquid nitrogen. The process of preservation is not labour-intensive but the cost of refrigeration equipment is fairly high. Provision must be made to safeguard against mechanical and electrical failure and the level of liquid nitrogen in the refrigerator must be maintained. Failure to replenish nitrogen stocks may result in the loss of an entire collection. Liquid nitrogen is hazardous and it is important that the tank is located in a well-ventilated area to ensure that the atmosphere does not become depleted of oxygen. It should not be handled without protective clothing to prevent injury from burns and the risk of explosion from glass containers should be noted.

Freezing in glycerol and storage between $-50°C$ and $-70°C$ is useful for large- and small-scale storage. Cells suspended in glycerol can be deposited inside hollow glass beads and stored at $-70°C$. The method, as described in (1), has been used for a wide variety of bacteria. Each glass bead provides sufficient inoculum for a sub-culture. Serial sub-culture may be required to provide a large inoculum and for this reason the method may not be suitable for unstable organisms. On a larger scale if a standardized inoculum is required for a series of experiments, a bulk preparation of culture can be suspended in glycerol, divided into inoculum-sized aliquots and stored frozen.

This method may also be used for strains requiring a specific inoculum concentration to induce secondary metabolite activity. Bulk suspensions can be frozen whilst the density of an aliquot is determined. The inoculum concentration then may be adjusted to meet the requirements of the strain.

Freeze-drying has been used to preserve many different groups of microorganisms, but may be unsuitable for certain non-sporing fungi. The advantages of this method lie in the convenience of the storage requirements for the freeze-dried products and in maintenance of viability without further attention on storage. The capital cost is high, the freeze-drying process is labour intensive and losses in viability may occur. It is the method of choice for most culture collections and it lends itself to batch culture, is excellent for storage of stand-by stocks and for long-term storage of cultures.

Many bacteria survive by sub-culture for several years. The method is inexpensive, but is labour-intensive if frequent sub-culture is required. Contamination is a major risk and dehydration may occur. Loss of viability is a hazard and cultures are prone to loss of stability, the risk increasing with the number of sub-culture cycles. This method is most suitable for bacteria which replicate by cell division and cannot be recommended for mutated actinomycete or fungal species.

Sub-culture incorporating a 'single spore' isolation step can be a useful method for the maintenance of inherently unstable actinomycete and fungal species. This method is often used to maintain production strains when others have failed. It is dependent upon the ready availability of a reliable, high-throughput product assay.

The storage of agar slant cultures under a layer of sterile medicinal paraffin to reduce the metabolic rate of the microorganism is the least time-consuming and probably the most inexpensive method available. As the culture is vulnerable to contamination from repeated use and subject to breakage, it should be considered a short-term storage method and should be supplemented with an alternative method. It is suitable for many fungal and actinomycete species.

As with any other field of experimental laboratory work, the user is advised to gain first-hand experience before applying any of these techniques to cultures of importance.

3. METHODS

3.1 Desiccation

Desiccation involves removal of water from a culture and preventing its rehydration. This objective can be achieved in a number of ways and the principal means are described in this section.

3.1.1 *Storage on soil or sand*

(i) *Soil preparation*. Sift dried loamy friable unfertilized garden soil through a 2 mm mesh sieve and dispense it into 25 ml wide-mouth McCartney bottles to a depth of 25 mm. Plug the bottles with non-absorbent cotton wool and sterilize at 2.1 bar steam pressure for 30 min and store at room temperature until required. If a lower operating steam pressure is all that is available, autoclave the soil on at least two occasions on an extended autoclave cycle and cool between each cycle. Prior to use moisten the soil with up to 1 ml sterile water, stir to give a damp powdery soil and steam for about 60 min.

(ii) *Inoculum*. Add a volume of sterile distilled water to a culture on agar and gently scrape the surface of the agar with an inoculating loop or wide-ended pipette to release spores. If the organism does not sporulate, a suspension of mycelium may be used. Remove the suspension from the agar by pipette and inoculate the soil preparation with 2 ml of suspension to provide around 2×10^7 colony forming units (c.f.u.) in the soil preparation. Do not exceed this volume; the soil must not become soggy because the organism will not grow through water-logged soil.

(iii) *Procedure*. Grow the organisms in the soil for a period of $10-14$ days at $24\,°C$. If the organism is growing poorly, inoculate the moistened steamed soil with a portion of the organism from a culture growing on agar and incubate for 10 to 14 days. After incubation, dry the soil by vacuum desiccation. This can be done in the chamber of a freeze-drier such as the Edwards EF03 or by desiccation over phosphorus pentoxide. If a desiccator is used place the bottles in the desiccator over a dish of phosphorus pentoxide and attach it to a pumping unit capable of removing 0.5 ml water min^{-1}. The pressure will fall rapidly and reach 1 mm of mercury, causing the cultures to foam up and freeze solid. The pressure will continue to drop and reach a pressure of 0.1 mm mercury. Renew the phosphorus pentoxide until moisture is no longer drawn off.

(iv)   *Storage.* When drying is completed cover the soil bottles with greaseproof or similar paper and store them in an airtight container at 4°C over calcium chloride or indicator silica gel.

(v)   *Recovery.* Revive the culture by sprinkling soil on to an appropriate medium and incubating it under suitable conditions for the organism. A pre-sterilized metal scoop which holds approximately 150 mg of soil is suitable for transferring the soil to the agar.

(vi)   *Sand.* Dispense acid-washed sand (40 − 100 mesh; BDH Chemicals Ltd) into 25 ml McCartney bottles to a depth of 25 mm, plug with non-absorbent cotton wool and sterilize by dry heat at 160°C for 2 h. Inoculate the sand with the suspension of culture and desiccate immediately.

3.1.2 *Storage in silica gel*

(i)   *Gel preparation.* Transfer purified non-indicator silica gel (mesh 6-22; BDH Chemicals Ltd) to 25 ml McCartney bottles to a depth of 25 mm and plug with non-absorbent cotton wool. Sterilize by dry heat at 170°C for 120 min and store in a warm dry atmosphere. To follow the method described in (2) the gels and harvesting medium should be chilled for 24 h before use and held in an ice bath during the inoculation stage.

(ii)   *Suspending medium.* Disperse 5 g of skimmed milk powder in 100 ml of distilled water and sterilize by autoclaving at 120°C for 20 min.

(iii) *Inoculation procedure.* Add chilled suspending medium to a culture on agar and gently scrape the surface of the culture to release spores, mycelium or cells. Transfer approximately 0.5 ml of the suspension to each silica gel bottle cooled in an ice bath. Leave the gels in the ice bath until the ice melts then remove them to room temperature. Leave for 10 to 14 days at room temperature until the gel crystals separate.

(iv)   *Storage and recovery.* Revive the culture by sprinkling gel crystals on to an appropriate medium and incubating under suitable conditions for the organism. A pre-sterilized metal scoop which holds approximately 150 mg is suitable for the transfer.

3.1.3 *Storage on paper strips*

(i)   *Paper preparation.* Cut Whatman Number 4 filter paper into 1 cm squares and sterilize in an aluminium foil envelope at 120°C for 20 min.

(ii)   *Inoculation procedure.* Follow the method in Section 3.1.1 (ii) to prepare a dense suspension of cells. Harvest the cells in a suspending medium, as detailed in Section 3.1.2 (iii). Use sterile forceps to transfer the paper squares to a sterile Petri dish. Moisten the paper squares with the suspension of cells. Drain excess moisture and return the paper strips to the foil envelope. Dry the contents of the envelope over phosphorous pentoxide for $2-3$ weeks, changing the desiccant as it becomes wetted.

(iii)   *Storage and recovery.* Store the foil envelopes over anhydrous calcium chloride or indicator silica-gel at 4°C. Use sterile forceps to remove a paper square from the foil packet and streak it across an agar plate of medium. Incubate the plate appropriately.

3.1.4 *Storage on gelatin discs*

(i)   *Preparation and inoculation procedure.* Follow the method in Section 3.1.1 (ii) to prepare a dense suspension of cells. Use nutrient broth as suspending medium. Add 10 g of gelatin to 100 ml of distilled water and dispense into 5 ml aliquots; autoclave at 120°C for 20 min and then maintain at 40°C. Add 1.0 ml of nutrient broth suspension to each gelatin bottle.
  Dispense droplets of the molten suspension into sterile Petri dishes and allow them to solidify. Freeze the droplet discs and then desiccate them under vacuum following the method in Section 3.1.1 (iii). The gelatin may be supplemented with 2.5% w/v nutrient broth powder and 5% w/v meso-inositol.

(ii)   *Storage and recovery.* Dispense anhydrous calcium chloride or indicator silica-gel into screw-neck vials to a depth of 25 mm and sterilize by dry heat at 190°C for 90 min. Transfer the dried discs to the vials and store at 4°C. To revive the culture dissolve a disc in nutrient broth and inoculate a solid medium with the suspension.

3.1.5 *Freeze-drying (lyophilization) — centrifugal*

Freeze-drying is a process in which water is removed by evaporation from a sample of frozen culture.

(i)   *Freezing procedure.* A freeze-drier such as the Edwards EF03 consists of two parts, the lower housing the refrigerator and its controls and the upper containing pump, centrifuge, manifold, Pirani gauges and control taps. The vacuum achieved is directly related to the vapour pressure of the water vapour at the temperature of the condenser and a good vacuum cannot be obtained unless this temperature is below -50°C and preferably nearer -70°C. The freezing results from rapid cooling caused by water boiling off at reduced pressure. To avoid foaming and to increase the surface area/volume ratio, the ampoules are centrifuged during the first stage of the process. Failure to cool rapidly enough results in a 'glassy' product and failure to centrifuge results in foaming causing lost or dirty plugs and soiled tubes.
  On completion of the primary drying the ampoules are constricted to facilitate sealing. This stage should be completed quickly to avoid exposure of the ampoules to air. Constriction is carried out on an extruder, such as the Edwards ampoule constrictor model number 2. The extruder consists of an air/gas burner pre-set

either to pilot-level or to a working level and positioned over rollers. Secondary drying takes place on a manifold and is usually completed overnight. In the event of not achieving a satisfactory vacuum, a high-frequency spark tester, such as an Edwards ST4M Spark tester, can be used to detect the position of the leak on the manifold. The Spark tester demonstrates a satisfactory vacuum by producing a pale blue glow within the ampoule, whereas a poor vacuum is indicated by a deep purple glow or no discharge. The tester should be used with care as excessive use can result in puncture of the ampoule.

After completion of the drying process the ampoules are heat-sealed *in situ* under vacuum with a flame guard protecting adjacent ampoules.

(ii) *Ampoule preparation.* Neutral glass freeze-drying ampoules (capacity 0.5 ml, SAMCO) should be plugged loosely with non-absorbent cotton wool. Prepare a plug approximately 18 mm long and insert it in the tube to leave $4-8$ mm protruding. Cover the tubes with foil and sterilize by dry heat at 160°C for 2 h. Do not char the plug or even allow it to become pale brown as this can reduce the viability of the subsequent preparation.

To label the ampoule prior to sterilizing, type or write the identification number on filter paper or blotting paper measuring 5 × 30 mm, insert it into the ampoule and then plug with cotton wool. If the label is to be placed on the outside of the ampoule prepare a label measuring not more than 5 × 23 mm and attach it with clear adhesive tape 30 mm from the rounded end of the ampoule. Do not put the label higher than this level or make it larger than the stated dimensions, or charring will occur at the extrusion stage.

(iii) *Suspending media.* A list of cryoprotectant suspending fluids appears in *Table 1*.

(iv) *Inoculation procedure.* The culture should be well grown and in an easily accessible form. Late exponential growth is often the optimal stage for freeze-drying. The culture may be grown either on agar plates, on the surface of 5 ml of slanted agar in a 152 mm × 15 mm tube, or in liquid culture. Liquid-grown cultures should be centrifuged to concentrate the cells. Add 5 ml of suspending fluid to each 152 mm × 15 mm test-tube or agar plate, gently rub off growth with a Pasteur pipette and dispense 0.2 ml of the suspension in to each tube. If a liquid culture has been used add 5 ml of suspending fluid to the centrifuged pellet of cells; resuspend the cells and dispense 0.2 ml into each tube. Take care to avoid creating aerosols and avoid contaminating the outside of the tubes. After filling, flame the tops of the tubes and replace the cotton wool plugs.

Ensure that the whole procedure is performed rapidly and is uninterrupted to the point of drying, otherwise dormancy of the culture will be lost. If interruption is unavoidable store the suspensions and suspending medium at 4°C during this interval. Incubate residual suspensions and suspending media at 30°C overnight to check for contamination.

Table 1. Suspending media for microorganisms.

Bacteria	*Yeasts*	*Fungi*	*Fungi, actinomycetes, bacteria*
(a) *Inositol serum*	*Glucose serum*	*Inositol milk*	*Tryptone yeast serum*
meso-Inositol 5 g	Glucose 7.5 g	Inositol 5 g	Tryptone 5 g
Horse serum No. 3	Inactivated horse serum No. 5	Skimmed milk 10 g	Yeast 3 g
Wellcome to 100 ml	Wellcome to 100 ml	Distilled water to 100 ml	Distilled water to 1000 ml
Sterilize by filtration	Sterilize by filtration	Autoclave 120°C for 20 min	Autoclave 120°C for 20 min
			Horse serum:
(b) Inositol broth			Difco 9207-65
meso-Inositol 5 g			Mix 2 parts horse serum to
Nutrient broth powder			1 part tryptone yeast
No. 2 Oxoid 2.5 g			
Distilled water to 100 ml			
Autoclave 120°C for 20 min			

(v)   *Drying procedure*

    (a)   Primary drying. Load the tubes into the centrifuge carrier and lower it onto the spindle in the centrifuge chamber, replace the perspex dome, switch on the centrifuge and spin for 60 min. Once the carrier is rotating switch on the vacuum pump to evacuate the chamber. This stage is completed after about $8-12$ h and is indicated by a Pirani gauge reading of 8×10^{-2} Torr.

        To accelerate the primary drying stage, remove the cotton wool plugs from the ampoules. Replace the cotton wool plugs with a cover of sterile gauze over each ampoule or group of ampoules. Replace the gauze with sterile cotton wool plugs on completion of the primary drying. Cut off the cotton wool protruding from the tube and push the remaining portion of the plug to approximately halfway down the ampoule.

    (b)   *Ampoule constriction.* Remove an ampoule of each culture type for viability testing. Switch the burner to a pilot flame and place the ampoules on to the rollers with the rounded end to the right-hand side and against the end stop; switch on the rollers. Turn up the burner to produce the working flame. As the tube becomes hot, it starts to elongate. Lower the flame to a pilot flame before the sides of the tube collapse. Do not seal the tubes at this stage. If one is sealed, exchange it with the culture set aside for viability testing.

    (c)   *Manifold drying.* Attach the extruded ampoules to nipples on a flat manifold, fill spare places with blank ampoules, switch on the vacuum pump, close the air inlet valve and open all valves to the manifold. Leave the ampoules on the manifold for at least 3 h (or overnight) at a vacuum of at least 10^{-1} Torr which should be achieved within 15 min of switching on the vacuum pump.

    (d)   *Sealing off.* Adjust the gas air mix of the sealing torch to produce a flame with an inner cone of 25 mm. Insert a flame guard between rows of ampoules and concentrate the flame on the ampoule constriction. Leave a rounded end on both the ampoule and the stub. Do not lose vacuum during this procedure; discard ampoules that have been sealed under low vacuum.

(vi) *Recovery.* Open an ampoule by making a score mark with a glass-cutter near the base of the cotton wool plug. Rest the ampoule in a bed of paper tissue and open by exerting pressure on both halves of the ampoule whilst holding the ampoule score side outwards. Remove the cotton wool plug and flame the neck of the ampoule, tap out the contents on to agar or gently add 0.1 ml of nutrient broth from a Pasteur pipette, draw up the broth into the pipette and release until the pellet of cells becomes suspended in the broth and transfer the suspension to a growth medium.

3.1.6 *Freeze-drying (lyophilization) – shelf-drying*

In shelf-drying the samples are pre-frozen before a vacuum is applied. The sample bottles are covered with gauze and frozen in a freezer or in a bath of solid carbon dioxide and acetone. The sample bottles are rotated in the bath to coat the inside wall of the container with a layer of frozen suspension and then placed in the chamber of a freeze drier. When dry, the containers are sealed to prevent rehydration of the sample.

3.2 **Freezing**

During freezing water becomes unavailable to the cells as it becomes frozen. The frozen cells are stored at low temperatures.

3.2.1 *Storage in 20% v/v glycerol at −50°C*

(i) *Preparation.* Add one part of glycerol to four parts of distilled water and mix. Dispense the glycerol into aliquots of 100 ml or less and autoclave at 120°C for 20 min. Small preparations of cells are preferable to larger preparations, because repeated freezing, thawing and transfer from a single culture will increase risk of contamination and may result in deterioration of the strain.

(ii) *Inoculation procedure.* Follow the method in Section 3.1.1 (ii) to prepare a dense suspension of spores, mycelium or cells. Use 20% v/v glycerol as suspending medium. Dispense as 1 ml aliquots into 2 ml polypropylene microtubes or as larger volumes in suitable containers. Store in a freezer at −50°C. A portion of a culture on agar may be submerged in 20% v/v glycerol and frozen.

(iii) *Storage.* Store the ampoules in self-sealing polythene bags or in polycarbonate boxes such as Nalgene cryoware boxes. These boxes have a numerical grid system on the outside of the lid to facilitate culture identification. Devise a filing system to assist rapid identification of a culture in the freezer.

(iv) *Recovery.* Bring an ampoule to room temperature and allow it to thaw. Transfer the thawed suspension to an appropriate medium and incubate.

3.2.2 *Storage in glycerol in glass beads*

The methods described in (1) and (3) provide novel approaches to storage of frozen suspensions in a cryoprotectant.

(i) *Preparation.* Wash 2 mm glass embroidery beads by the following procedure: (a) detergent, (b) dilute hydrochloric acid, (c) tap water to return to a neutral pH, (d) distilled water. Dry at 45°C. Dispense the beads 20−30 to a 2 ml screwcap glass bottle and sterilize by autoclaving at 120°C for 20 min.

(ii) *Suspending medium.* The composition of the suspending medium for aerobic bacteria appears in *Table 2*.

(iii) *Inoculation procedure.* Follow the method in Section 3.1.1 (ii) to prepare a dense

Table 2. Suspending medium for aerobic bacteria.

Bacteriological peptone (Oxoid L34)	8.0 g
Beef extract (Oxoid Lab-Lemco L29)	8.0 g
Sodium chloride	5.0 g
Glycerol	200 ml
Distilled water to pH 7.3−7.4	1000 ml

Sterilize by autoclaving at 120°C for 20 min

suspension of spores, mycelium, or cells from a non-selective medium. Dispense the suspension into vials containing beads. Wet the beads thoroughly and remove excess suspension from the vial.

(iv)   *Storage and recovery*. Place the vials in trays and store in a freezer at −70°C. Freeze the vials on their side to facilitate removal of a single bead later. Revive the culture by removing a bead from the vial with a sterile micro-spatula, rub the bead over the surface of a solid medium or wash it in nutrient broth and use the suspension to inoculate solid medium. Replace the vial in the freezer immediately to prevent the remaining contents from thawing.

3.2.3 *Storage in liquid nitrogen*

(i)   *Preparation*. Add one part glycerol to nine parts distilled water and mix. Dispense the glycerol into aliquots of 100 ml or less and autoclave at 120°C for 20 min.

(ii)   *Inoculation procedure*. Follow the method in Section 3.1.1 (ii) to prepare a dense suspension of spores, mycelium or cells. Use 10% v/v glycerol as suspending medium. Dispense 1 ml aliquots into sterile 2 ml polypropylene tubes and seal the tube with an internal threaded cap. Store the ampoules at 4°C for an hour and then suspend the ampoules in the vapour phase of a liquid nitrogen refrigerator. To monitor the rate of freezing introduce a thermocouple probe through a drill hole in an ampoule cap and attach it to a chart recorder. Raise or lower the freezing position in the tank to decrease or increase the rate of freezing. A freezing rate of $1°C\ min^{-1}$ is best for most organisms.

Polypropylene drinking straws can be used as an inexpensive alternative to polypropylene ampoules. Cut the straws into 2.5 cm lengths and seal at one end in a Bunsen burner flame. Place the straws in a glass Petri dish and autoclave at 120°C for 20 min. Remove as required from the dish using sterile forceps and fill with inoculum suspended in cryoprotectant using a Pasteur pipette and then seal the open end in a flame. Store up to six straws in a 2 ml screw-capped ampoule.

(iii)   *Storage*. Slot the frozen ampoules into aluminium canes and sleeve the canes in PVC to prevent ampoule loss. Store the canes in canisters and hang them over grooved cross-struts in the liquid nitrogen tank or store them in polycarbonate boxes as described in Section 3.2.1 (iii). The boxes have a number of small holes punched in them to allow escape of liquid nitrogen and can be placed in stainless steel racks designed for this purpose and stored in the tank.

3.3 **Storage by sub-culture**

This method consists of inoculation of a culture on to a suitable medium contained in a tube or bottle, incubation to achieve growth and storage. The process is repeated at intervals to ensure the preparation of a fresh culture before the old one dies. It is common practice to attempt to simulate the natural conditions of the organism.

3.3.1 *On agar in the refrigerator*

Sub-culture the organism and store in a refrigerator at 4°C.

3.3.2 *On agar at room temperature*

Sub-culture the organisms and store at room temperature.

3.3.3 *On agar blocks at room temperature*

(i)   *Preparation.* Cut an agar block from the growing edge of a culture and transfer it to sterile distilled water in a McCartney bottle. Screw down the lids and store at room temperature.

(ii)   *Recovery.* Transfer the agar block to fresh agar and incubate.

3.3.4 *On agar under sterile mineral oil*

(i)   *Preparation.* Inoculate a culture onto the surface of 5 ml slanted agar in a 152 mm × 15 mm tube or 10 ml bottle, plug with sterile non-absorbent cotton wool and incubate the culture. Sterilize medicinal liquid paraffin at 140°C for 3 h and cool. Cover the surface of the mature culture with the liquid paraffin to a depth 1 cm above the top of the agar. Store the oiled slant at room temperature or at 4°C.

(ii)   *Recovery.* Remove a small amount of culture on a flat-ended incurved sterile needle. Drain off as much oil as possible and streak the culture on to solid medium. The growth rate may be impaired due to the remaining oil, and often more than one sub-culture may be required to recover the organism.

3.3.5 *Sub-culture incorporating a single spore selection step*

Strains which have been subjected to a mutation programme are frequently found to be unstable. Maintenance that incorporates a 'single spore' purification step is more suitable for these organisms than serial sub-culture.

Preparation. Follow the method in Section 3.1.1 (ii) to prepare a dense suspension of spores. Use distilled water as suspending fluid or distilled water and a wetting agent such as 0.1% Tween 80 or Triton X100 if the spores are hydrophobic. Transfer the suspension to a 25 ml wide-mouth McCartney bottle containing around 16 sterile spherical glass beads diameter 5 mm, shake or vortex the suspension to break up clumps of spores and filter the suspension through a bed of sterile glass wool retained in a filter funnel to remove debris and mycelium. Collect the filtrate in a sterile test-tube and filter it through a membrane filter. Selection of the appropriate filter depends on the dimensions of the spores; the filter should exclude all but single spores. Prepare a dilution series of the suspension in water and spread 0.1 ml aliquots of the dilution series on to solid agar. By plating a range of dilutions a count of 20−50 colonies per plate on one dilution will be achieved. Incubate the plates. After incubation transfer a colony on a sterile wooden cocktail stick into a wide-mouth McCartney bottle containing glass beads and 2 ml of sterile water or 1/4 strength Ringer's solution. Shake the bottles to give a suspension of culture and inoculate the surface of an agar slant in a 150 mm × 25 mm tube and incubate. Several slants can be prepared from each suspension. Screen one slant of each isolate in the product assay and retain the highest-yielding isolates as master cultures. Sub-culture these masters as required to provide batches of working cultures. Retain all mature cultures at 4°C to give a shelf life of at least 4 weeks. Retain a 'first generation' slant of the highest-yielding isolate as a parent strain for the next single-spore isolation procedure.

4. CULTURE MANAGEMENT

4.1 Post-preservation testing

4.1.1 *Viability, stability, purity and productivity testing*
The cultures should be checked for viability, purity and retention of desired characteristics after preservation. In general a simple growth test for viability will be sufficient and this test may also be used to check for contaminating organisms. It is advisable to monitor the purity of residues of suspensions used for preservation. If an indication of survival rate is required a counting method should be used. If pre- and post-preservation viability counts are required, remove a measured volume from the initial suspension, prepare a dilution series and plate a measured volume on to agar. To calculate the number of colony-forming units in the suspension, count the number of colonies on a plate and multiply this figure by the dilution factor. Prepare and plate a similar dilution series on the revived culture and compare the two results. Post-preservation activity screens should be performed on organisms producing secondary metabolites. A single revived culture can be used for all post-preservation testing.

4.2 Information storage and retrieval

4.2.1. *Culture records and computer management*
A culture should be given an identification unique for that organism; elaborate systems should be avoided. The culture should be labelled with this identification rather than by abbreviated codes needing a cross-reference. A 12-digit number can be fitted on to a label measuring 5 × 30 mm. Write labels clearly in waterproof indelible ink. A documentation system is an essential accompaniment to culture maintenance and a strain-record should include the following information:

(a)   'In-house' collection and reference number.

(b)   Source and source reference number, or equivalent number in other collections.
(c)   Organism name or synonym.
(d)   Alternative states.
(e)   Classification.
(f)   Date of deposit.
(g)   Known products and yields.
(h)   Pathogenicity/licensing status.
(i)   Cultural requirements: medium, temperature for growth, pH, and nutritional requirements.
(j)   Isolation media and method.
(k)   Culture description on a range of media.
(l)   Culture use, if any.
(m)   Maintenance method.

Computers are ideally suited to culture collection management. Strain data can be stored and a programme can be devised to allow rapid information retrieval. A separate 'ISSUES' file may be kept to record the origin and performance of each culture issued from the collection. A card index filing system containing strain information continues to have a use within a computerized system. The card should accompany a culture through the preservation pipeline. The information on the card is used for post-freeze-drying observations and alerts the user to organisms demanding special attention.

4.3 Culture protection from contaminating organisms

4.3.1 *Mite prevention*
Mites can devastate culture collections. They may destroy the culture by eating it, or may move between cultures carrying spores on their bodies causing cross-contamination. Work surfaces must be kept clean and be washed from time to time with acaricides such as Jeyes fluid or Tegadon. Cultures containing mites should be sealed and auto-claved. Tractor vaporizing oil (TVO) acts as a mite deterrent and can be placed in dishes around the laboratory.

4.4 Safety

Eye protection should always be worn whilst evacuating equipment and handling glass under vaccuum, and full face protection worn for liquid nitrogen. It is good practice to wear eye protection for all laboratory manipulations. Additional precautions should be taken when working with:

4.4.1 *Pathogens*
Pathogens should be handled with caution at all times. Cultures for freeze-drying should always be plugged with cotton-wool and pre-frozen prior to transfer to the drier. Ampoules should be filled and opened in an exhaust-protected cabinet, and all glassware including Pasteur pipettes should be soaked in 1% chloros or equivalent before disposal. Cultures should be autoclaved prior to disposal.
    When pathogens are stored in liquid nitrogen they should never be stored in glass containers because of the explosion risk.

4.4.2 *Desiccant-phosphorus pentoxide*

Phosphorus pentoxide is corrosive and should be handled in a fume cupboard. A protective face mask and gloves should be worn whilst dispensing the dry powder. Spent phosphorus pentoxide is best left to hydrate by exposure to air, preferably in a fume cupboard. The sludge should be dissolved in water and with care washed to drain.

4.4.3 *Liquid nitrogen*

In addition to full face protection, protective gauntlets should be worn when handling liquid nitrogen. Thawing samples should be shielded until they reach room temperature, since pressure build-up can occasionally cause explosions. The storage vessel should be located in a well-ventilated area fitted with an extraction fan.

4.4.4 *Vacuum equipment*

Equipment under vacuum may fail by implosion. Laboratory staff should be protected by having all desiccators under vacuum in wire cages. Glass manifolds are covered in plastic webbing and are kept behind rigid perspex screens.

4.5 Service collections

A number of service collections and federations exist throughout the world. In addition to their main functions of collecting, maintaining and supplying cultures, they provide a number of other services including culture identification and preservation, safe-deposit facilities and depositories for patent purposes.

4.5.1 *Main collections*

A list of the main collections appears in the appendix (p. 221).

5. INOCULUM PREPARATION

The organisms used for secondary metabolite production are generally fungi and actinomycetes. The fungi are eukaryotes with a complex nucleus of several chromosomes enclosed in a nuclear membrane, whilst the prokaryotic actinomycetes have a single unenclosed circular chromosome. Despite the difference in their nuclear organization, cell-wall composition and cellular organization, producer-strains from either group are often handled in a similar way.

For all organisms, the general principle is to minimize the number of 'generations' that a culture goes through from starting culture to the final harvesting of a fermentation broth. Mutated strains are often genetically unstable, and low-yielding variants may appear within the population; on sub-culture these may out-grow the parental strain, resulting in lost productivity. On agar, strain deterioration may be manifested as morphological variants within a culture, and plating out of a culture at the end of a fermentation will indicate the extent of instability within a strain.

To minimize the problems of a strain deterioration, all cultures preserved as master cultures for production purposes should be prepared from 'single-spore' isolations of the parental strain. The method given in Section 3.3.5 should be followed, prior to preservation of the culture by the chosen method.

A survey of secondary metabolite yield from a population of 'single-spore' isolates would usually include within the population a group of isolates with yields lower than the parent strain, a group of isolates with yields higher than the parent and a majority of isolates with yields at a similar level to the parent. It is important that the proportion of low-yielding isolates in a population is monitored and excluded from the population used for fermentation. The methods outlined in the following section assume that this attention is given to the strain and that starting strains are considered to be 'single-spore' isolates.

5.1 Profile of inoculum development

The process of inoculum development is outlined in *Figure 1*.

5.1.1 *Preparation of working cultures*

A 'first-generation' culture on agar is prepared from the preserved stocks by the methods set out in Sections 3.1−3.3. This culture is normally sub-cultured to provide a set of working cultures. If the fermentation is to be small scale, this 'first-generation' culture may be used.

The organism should be inoculated over the entire surface of the agar medium, otherwise growth will be uneven and a uniform culture will not be achieved. The individual colonies will spread over only a limited distance, therefore it is better to use a suspension of culture rather than a dry transfer of inoculum.

A dense suspension of spores and mycelium harvested from the surface of 15 ml of slanted agar in a 152 mm × 25 mm tube is usually an adequate inoculum for 50−100 ml of inoculum medium. If larger volumes of spore suspension inoculum are required, 45 ml of slanted agar in a 200 ml flat-sided medicine bottle provides a larger surface area for culture growth.

Preparation. Add a volume of sterile distilled water to a culture on agar and gently scrape the surface of the agar with an inoculating loop or wide end pipette to prepare a dense suspension of cells. Transfer 0.1 ml of this suspension to the surface of the

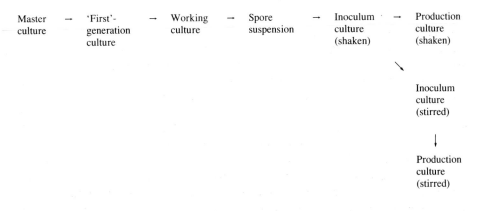

Figure 1. Inoculum development process.

agar. Use correspondingly larger volumes of suspension for larger volumes of agar. After all the working slants have been set up, inoculate the primary culture with 5 ml nutrient broth. Incubate at 30°C and 37°C for 24 or 48 h and inspect the broth for contamination. Incubate the slants with the agar surface horizontal so that the liquid soaks evenly into the surface of the agar to encourage uniform sporulation. At maturity transfer the working cultures to 4°C until required for use. For most strains, working cultures should be renewed at monthly intervals.

5.1.2 *Preparation of a spore suspension*

Most streptomycetes and many fungi sporulate. The spores usually arise in chains and can readily be separated by shaking them in water.

Procedure. Add sterile water to a culture on agar to give an approximate ratio of 9 ml water to 15 ml of slanted agar. Scrape the surface of the culture with an inoculating loop or wide-end pipette, first with gentle pressure and then more vigorously. Transfer the suspension by pipette to the inoculum vessel or flame the necks of the containers and pour the suspension into the inoculum vessel.

If an experiment involves setting up a series of inoculum flasks from a given strain it is better to transfer the suspensions from a number of working cultures into a sterile flask and divide the bulked inoculum between the series of flasks rather than to inoculate each flask with a similar though separate inoculum suspension. If the fermentation incorporates a two-stage inoculum, the bulking of the inoculum will automatically occur at the secondary inoculum stage.

Inoculate tubes or flasks of nutrient broth with an aliquot of suspension. Incubate at 30°C and 37°C and inspect for contamination after 24−48 h.

5.1.3 *Preparation of inoculum*

If good fermentation results are to be obtained optimal growth conditions are essential and the fermentation must be inoculated with a culture in perfect condition at the correct density. The effect of inoculum concentration and quality on final yield should be explored and the inoculum density adjusted to accommodate the requirement. This topic has recently been reviewed (4). Shake-flask fermentations are useful for this type of experiment. They are usually run with 50 ml medium in 500 ml Erlenmeyer flasks and shaken at 250 r.p.m. on a rotary shaker with a 50 mm orbit. Shake-flask experiments set up under these conditions can be used to explore many aspects of inoculum preparation.

(i) *Optimizaton of spore concentration in the inoculum.* Prepare a suspension of cells as described in Section 5.1.1. Harvest the cells in the smallest possible volume to prepare a dense suspension. Remove a measured amount from the suspension and dilute it in known volumes of sterile water to cover a range of concentrations. Inoculate 5 ml from each concentration of spores into an accurately measured volume of 50 ml of inoculum medium in a flask, shake until growth is sufficiently dense to give approximately 80% cells on settling (this growth time should be established in advance as inoculum should be transferred quickly to prevent cell death), then transfer a 10% (v/v) inoculum into the production flask. Harvest the flasks at the end of the production time and assay for product. Replicate flasks should be screened. Remove an 0.1 ml aliquot from the residual suspension of each concentration used in the experiment and spread it on an agar plate using a bent glass rod sterilized by flaming in alcohol. Incubate the plates and calculate the concentration of c.f.u. in each suspension following the method described in Section 4.1.1.

The results from this experiment will establish the spore concentration range required to optimize secondary metabolite production and will indicate the working slant requirement for a fermentation.

(ii) *Preparation of one- and two-stage inoculum cultures.* Slant cultures are used to prepare spore suspensions which are used to start shake-flask cultures. The shake-flasks provide a quantity of mycelial growth which is used to start either the fermentation in shake-flasks or to start the secondary inoculum culture in a fermenter. The method used to prepare a spore suspension for a one-stage or a two-stage inoculum is the same. Spore suspensions are prepared following the method in Section 5.1.2. The only difference is in the scale of preparation. Shake-flask fermentations are usually operated in 50 ml volumes whilst processes in fermenters may vary from 2 to 5000 l. This scale of operation involves a corresponding scale-up of inoculum.

To provide an inoculum for a larger volume fermentation a two-stage inoculum is used. The first stage is usually shaken, whilst the second stage is prepared as a fermenter-stirred culture. To provide the volumes of inocula required, the first-stage inoculum is often prepared as 500 ml of culture in 2 l flasks. The number of flasks is determined by the volume of the secondary inoculum, which in turn is determined by the working capacity of the fermentation vessel. Retain an aliquot of inoculum medium, pre-inoculation suspension and mature inoculum. Inoculate nutrient broth with the samples and incubate at 30°C and 37°C for 24−48 h and inspect the broths for contamination.

6. SCALE-UP OF THE INOCULUM

6.1 Introduction

A fermentation production process requires biomass which is highly efficient and predictable in performance. In order to satisfy these criteria an inoculum of consistent quality and sufficient volume is required. The development of the inoculum, through the various stages, terminating in the inoculation of the production fermenter, has a definite effect on the subsequent performance of the process−in some cases bad management of the inoculum can result in an almost complete loss of productivity.

In laboratory research it is also imperative, when attempting to optimize a process, that a reproducible, reliable baseline condition is achieved; without such a baseline any increases in yield/productivity are difficult to assess.

The simplest system of inoculum procedure uses vegetative seed stages on similar medium all the way through the stages from shake flask to production fermenter. This implies vigorous growth throughout the stages; the medium should not become exhausted or highly unbalanced prior to the transfer to the next stage; the desirable amount of inoculum is at least 5% of the next volume. In a laboratory fermentation the inoculum stages rarely exceed two stages, whereas as many as six may be required for certain production processes.

When using shake flasks as a source of inocula for laboratory fermenters, it is very difficult to obtain on-line information regarding the physiological state of the micro-organisms contained within the culture. Dry cell weight, viable counts, etc., are of little use when the delay time involved in such analysis is considered. Without such information it is difficult to predict what part of the growth curve the inoculum is in prior to the inoculation of the production fermenter. The difficulty in identifying the physiological state of the microorganisms is made worse when using complex media (containing sparingly soluble components) where even visual estimation of growth is not possible.

The following sections describe the advantages of being able to control and monitor (on-line) the inoculum and consequently being able to optimize the time of transfer to the production fermenter in order to provide a predictable, reliable process; and also in certain cases to increase the productivity.

6.2 **Methodology**

In order to achieve control and on-line analysis of the inoculum the final stage must be carried out in a fermenter. Such a seed fermenter may only be occupied for one or two days a week, which is a luxury many researchers can ill afford! However, it has the advantage of being able to provide information, at a later date in the project, for the design of the production plant (if applicable) when the final stage of the inoculum will almost certainly be provided via a fermenter.

6.2.1 *Equipment*

(i)   *Fermenter.* The inoculum fermenter should be capable of providing at least three production vessels with a volume of inocula which is 10% of the working volume of the production vessel. The volume of the fermenter must also be large enough to be not significantly affected in terms of mass transfer, control, etc. when three such volumes have been removed.

(ii)   *Controls.* The fermenter must have control for temperature and pH (preferably acid and/or alkali) and be capable of monitoring dissolved oxygen tension (DOT).

(iii)   *Analysis.* Accurate exit fermenter gas analysis of oxygen and, in particular, carbon dioxide is essential for the development of the inoculum as described in this chapter. The most efficient and accurate method available for such analysis is via a mass spectrometer (the MM 8-80 from VG Gas Analysis Ltd is recommended). This type of analyser has the advantages of automatic calibration, rapid sample time, very little drift, and easy interfacing to a computer. They are expensive, and alternative analysers such as the infrared type for carbon dioxide analysis and paramagnetic for oxygen analysis are cheaper although not as reliable.

    A method of measuring the exit gas flow rate is also required. A relatively cheap and accurate method is via a rotary gas meter [Alexander Wright and Co. (Westminster) Ltd].

6.2.2 *Basic techniques*

On-line analysis which describes the growth of the inoculum must be readily available in numerical form in order to indicate the optimum time of transfer of the inoculum into the production fermenter.

Usually the most common criterion for assessing the quality of a vegetative inoculum is the concentration of biomass present in the culture. Ideally one would measure the extent of growth in the culture and transfer at an appropriate point along the growth profile (5). Measurements such as turbidity, dry cell weight, packed cell volume, pH, dissolved oxygen and morphological form have all been used with varying degrees of success (6).

The above methods, however, are either not on-line (e.g. dry cell weight, packed cell volume) or not reproducibly accurate enough in numerical terms (e.g. packed cell volume, morphology). Hence the success of fermenter gas analysis for this purpose; it is on this technique that the following protocol is based.

(i) *Preservation of microorganisms.* This has been comprehensively covered in Section 2.

(ii) *Development of inoculum growth medium.* The shake-flask medium for the inoculum stage is typically well buffered, containing high concentrations of phosphates and perhaps chalk and often including sparingly soluble components such as cotton-seed residues (e.g. Pharmamedia) and soya-based products (e.g. HISOY). Whilst this type of medium is obviously suitable for shake flasks it can be modified, although it is not a necessary step in the development process, in order to make it more suitable for the fermenter stage of the inoculum grow up. This stage will have pH control (see Section 6.2.2 (iv)) and consequently buffers are not required; the phosphate concentration can be reduced significantly and replaced by other phosphate salts, and chalk removed entirely from the medium (this also reduces the abrasiveness of the medium which will reduce the wear on a bottom-drive mechanical seal if applicable). The medium may require enriching with specific components in order to prevent exhaustion of such components from the medium during the growth of the organism in the inoculum grow up.

(iii) *Sterilization of inoculum medium.* The sterilization conditions of the media via autoclaving for shake flasks and autoclaving or *in situ* sterilization for fermenters should be kept constant throughout the development process. The composition of complex medium can be significantly altered by the conditions of sterilization. Complex nitrogen sources such as Pharmamedia and HISOY are solubilized by high temperatures (the higher the temperature the greater the degree of solubilization). The effect of this is to release amino acids and peptides into the medium, making the nitrogen more available to the microorganism and so altering the balance of the medium.

(iv) *Conditions for growth of the inoculum.* The growth conditions for the inoculum will almost certainly have been obtained from a reference paper, text book or colleague and the conditions will refer to shake flasks. These will need translating to satisfy the conditions in the fermenter stage of the process. The temperature will be the same for both stages; however, translation of pH, shaker speed, and inoculation time is more difficult (it is better to control the pH of the fermenter stage in order to increase the predictability of the growth kinetics rather than relying on the use of buffers).

The pH at which the fermenter is controlled is best estimated by following the pH profile of the primary shake flask via frequent sampling throughout the growth profile; from such a profile the pH which was prevalent for the major phase of the growth can be taken as the control value.

The translation of the agitation speed of the orbital incubator to that of the fermenter is more difficult. The simplest 'rule of thumb' is to use an agitation speed and aeration rate that maintains a positive dissolved oxygen tension throughout the course of the inoculum grow-up. An aeration rate of 0.5 v/v/min combined with an agitation speed of 500 r.p.m. is usually sufficient, in a baffled fermenter, to satisfy the above criterion; however, the conditions which best suit the process in question must be identified by the design of a small experimental programme varying the agitation and aeration rate of the inoculum fermenter and monitoring the dissolved oxygen tension.

The number of inoculum stages for small-scale laboratory production fermenters (i.e. with a volume less than 30 litres) will usually be two — a primary shake-flask stage and a secondary fermenter stage.

6.3 Development of an inoculation protocol

Once the medium and conditions have been obtained for the growth of the inoculum in both the shake flask and the fermenter the inoculation protocol can be developed and optimized, based on the following sequence:

$$
\begin{array}{ccccc}
 & 10\% \ (\text{v/v}) & & 10\% \ (\text{v/v}) & \\
\text{Primary inoculum} & \longrightarrow & \text{Secondary inoculum} & \longrightarrow & \text{Production} \\
 & & & & \\
\text{(Shake flask)} & & \text{(Fermenter)} & & \text{(Fermenter)}
\end{array}
$$

6.3.1 *Typical procedure*

(i) *Primary inoculum (shake flask).* Normally a mature slope of the microorganism is used to inoculate 300 ml of shake-flask medium contained in a 1000 ml non-baffled flask.

The incubation time for this stage is not too critical — however, once determined it must be adhered to in order to maintain the integrity of the process. Between 24 and 72 h is common. A 48 h inoculation period allows a purity sample to be removed after approximately 24 h and the results known prior to the transfer of the shake flask into the fermenter.

At the end of the incubation period the contents of the flask are transferred into 3 litres of fermenter inoculum media (i.e. 10% v/v) aseptically and with a minimum of delay. It is important, throughout the protocol, to keep all transfer times of inocula to a minimum — the viabilities of certain microorganisms, particularly *Actinomycetes*, decrease rapidly in conditions of zero agitation and aeration.

(ii) *Secondary inoculum (fermenter)*. The temperature, pH, aeration, and agitation have all been previously determined [see Section 6.2.2 (iv)] and set accordingly. Antifoam may have to be added to the culture throughout the fermentation, depending on the process.

The first fermentation is to generate data that can be used to define the inoculation protocol, in particular the carbon dioxide production rate (CPR). This rate is calculated using the percentage of CO_2 in the fermenter gas in the following equations and expressed in nmol l^{-1} min^{-1}.

(a) Conversion of ml min^{-1} at NTP to mM h^{-1}

1 g mol of gas at 0°C and 760 mm Hg occupies 22.412 litres

1 g mol of gas at 20°C and 760 mm Hg occupies:

$$22.412 \times \frac{293}{273} \text{ litres}$$

$$= 24.03 \text{ litres}$$

Thus 1 mmol at NTP occupies 24.03 ml

or 1 ml at NTP $= \dfrac{1}{24.03}$ mmol

Thus 1 ml min^{-1} at NTP $= \dfrac{1}{24.03}$ mmol min^{-1}

$$= 0.0416 \text{ mmol min}^{-1}$$

(b) Conversion of % CO_2 to mmol CO_2 l^{-1} min^{-1}

$$\frac{\% \ CO_2}{100} \times 0.0416 \times \frac{G_f}{v} = \text{mmol } CO_2 \text{ litre}^{-1} \text{ min}^{-1}$$

where G_f is the gas flow rate (ml min^{-1}) and v is the fermenter working volume (litres).

The fermentation is terminated when the CPR has fallen to a rate which is constant. During the time course discrete samples for secondary analysis (e.g. dry cell weight, residual substrates, pH, etc) can be taken − however, the volume of these samples must be kept to a minimum and the corrected volume reading taken and used in the calculation of CPR.

From the CPR profile (a typical one is shown in *Figure 2*) it can be decided at which points on the profile of the subsequent fermentation samples are to be removed from the inoculum fermenter and used to inoculate production fermenters. The points indicated in *Figure 2* are at suitable phases of growth for this initial experiment − early experimental, mid experimental and midway through the declining CPR period.

For the inoculation transfers a charge of 10% (v/v) is recommended (in many

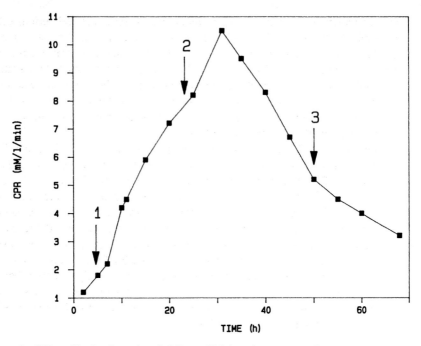

Figure 2. CPR profile showing points (1,2,3) at which inocula are removed.

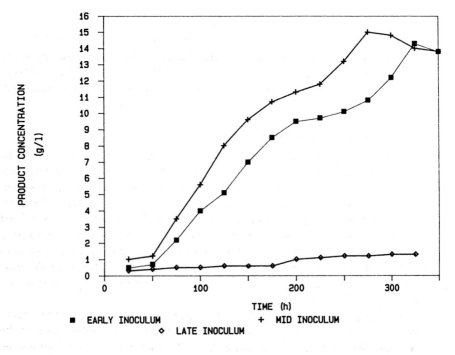

Figure 3. The effect of inoculum age on product formation in production fermenters.

cases significantly smaller volumes can be used without any deleterious effect on productivity – use of 10% (v/v) assumes a confident prediction of no loss in productivity). It is important to remeasure the inoculum fermenter volume each time an inoculum is withdrawn and use the mean volume in the CPR calculation. Again the transfer time of these inoculations must be kept to a minimum – no more than 2 min is generally acceptable. The conditions under which the production fermentations are run must remain constant throughout the development of the inoculation protocol.

Once the data from the first series of production fermentations has been compiled, the optimum transfer time of the secondary inoculum can be predicted in terms of CPR. It is of practical importance, especially when considering the scale-up of processes, to know how tolerant such a process is and, therefore, a range of CPRs over which the maximum productivity of the process is maintained. If this is required a second experiment must be carried out, choosing CPRs in varying degrees of proximity to the initial optimum rate, until such a range is obtained. Indeed, the initial optimum CPR for inoculation may not represent the best rate for maximum productivity, but merely be better than the other two rates chosen from profile. If this is the case the second series of experiments will provide not only the maximum productivity of the process but also the CPR range that will provide it.

6.4 Evaluation of the inoculation protocol

The main aim in developing a protocol of this type is to obtain reproducible data, thereby being able to confidently establish a baseline and then being able to optimize the process further. In certain cases the overall efficiency of the process can also be improved.

Some examples of different microorganism systems will be briefly described in the following section.

6.4.1 *A Streptomycete fermentation*

The first example is used to demonstrate the effect of transferring the inoculum at various points on the CPR profile. The product in this example is on animal growth promoter produced as a secondary metabolite in batch culture. *Figure 1* indicates the points at which the inocula have removed and transferred to the production fermenters. The subsequent yield and CPR profiles of the production fermenters are shown in *Figures 3* and *4* respectively. The effect on yield is very obvious – if the inoculum is removed from the fermenter after the CPR has peaked and is decreasing, a significant loss in productivity follows. The production CPR profiles suggest that this is not due to a decrease in the viability of the *Streptomycete* late in the growth curve but more associated with the physiology of the microorganism.

These data illustrate the problems that can be encountered, especially in secondary metabolite fermentations, when using shake-flask inocula; if the metabolism of the inocula in the shake flask is between points 2 and 3 (or after point 3) on *Figure 3* the final yield of the production fermenter may be low, variable and not reproducible.

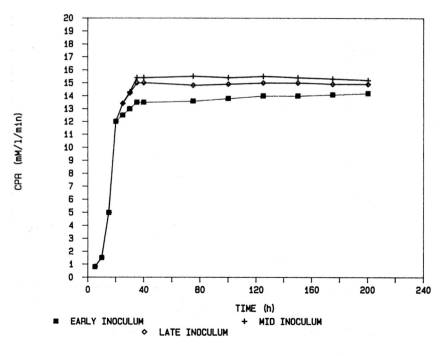

Figure 4. The effect of inoculum age on CPR in production fermenters.

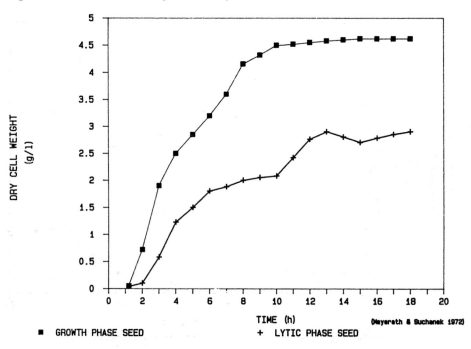

Figure 5. Optimization of biomass production by selection of the appropriate inoculum type.

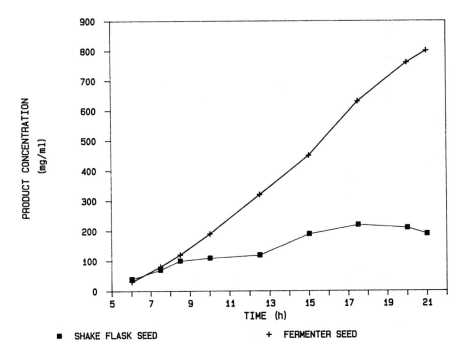

Figure 6. Increased secondary metabolite production due to the use of fermenter seed.

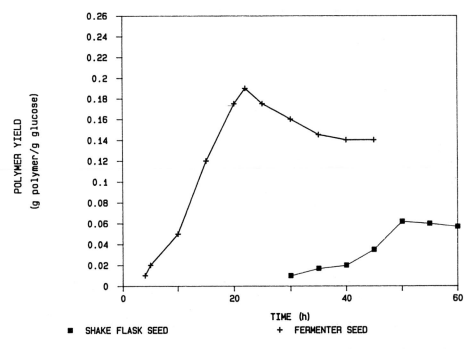

Figure 7. Time course of polymer producing fermentations. (The improvement in polymer yield and shortening of fermentation time when a fermenter derived inoculum is used are clear.)

6.4.2 *An Aspergillus fermentation*

In this example (see *Figure 5*) biomass production was being optimized and it was found that the culture that had been inoculated with an autolysing seed provided a greater concentration of mycelia throughout the fermentation than one seeded with an inoculum in the exponential phase of growth (7). Such a phenomenon is often encountered.

6.4.3 *A Cephalosporium fermentation*

The production of a secondary metabolite which is potentially active against specific types of cancer was significantly increased (approx. 300%) via the development of an optimized fermenter based inoculation protocol. These improvements can be seen in *Figure 6*.

6.4.4 *A bacterial polymer fermentation*

Figure 7 shows the effect of a fermenter inoculum when used in the batch production of a microbial cellulose polymer. In this example not only is the yield increased but the process is intensified—the maximum yield being obtained in approximately 30 h compared with 50 h via the shake-flask inoculated fermenter.

The above examples serve to illustrate the advantages of developing a fermenter based inoculation protocol. Not all processes will benefit from the increases in yield described but what such a protocol offers is reliable, baseline data and without such data process optimisation is very difficult to interpret.

7. ACKNOWLEDGEMENT

One of the authors (C.P.) would like to thank Mrs Beverley Hoult for typing the manuscript.

8. REFERENCES

1. Feltham,R.K.A., Power,A.K., Pell,P.A. and Sneath,P.H.A. (1978) *J. Appl. Bacteriol.,* **44**, 313.
2. Kirsop,B.E. (1984) In *Maintenance of Microorganisms.* Kirsop,B.E. and Snell,J.J. (eds), Academic Press, New York, p. 118.
3. Jones,D., Pell,P.A. and Sneath,P.H.A. (1984) In *Maintenance of Microorganisms.* Kirsop,B.E. and Snell,J.J. (eds), Academic Press, New York, p. 35.
4. Calam,C.T. (1987) In *Process Development in Antibiotic Fermentations.* Cambridge Studies in Biotechnology 4, Cambridge University Press.
5. Hockenhull,D.J.D. (1980) In *Fungal Biotechnology.* Smith,J.E., Berry,D.R. and Kristiansen,B. (eds), Academic Press, London, Vol. 3, p. 1.
6. Calam,C.T. (1979) *Folia Microbiol.,* **24**, 276.
7. Meyrath,J. and Suchaneck,G. (1972) In *Methods in Microbiology.* Morris,J.R. and Ribbons,D.W. (eds), Academic Press, London, Vol. 7B, p. 159.

CHAPTER 4

Fermentation modelling

C.G.SINCLAIR and D.CANTERO

1. INTRODUCTION

The purpose of fermentation modelling is to bring order to the mass of data which results from a practical fermentation experiment and to express the results in a concise form which is intelligible to colleagues and to those who wish to make use of the results. One of the principle uses of a fermentation model is to design large-scale fermentation processes using data obtained from small-scale fermentations.

As an example, let us take a simple aerobic yeast fermentation. Pure yeast culture is first transferred from a slope to a shake flask, grown in an orbital shaker for 12 h; the shake-flask contents are then aseptically added to a 5 litre laboratory fermenter. This is equipped with the usual agitator, air supply control equipment, probes for measuring temperature, pH, dissolved oxygen and, hopefully, off-gas analysis equipment. It will have been charged with a growth medium and be free of organisms prior to inoculation. A sample of the inoculum and of the medium will be taken initially (time $t = 0$), then, every 30 or 60 min, a sample of the fermenter contents will be taken for analysis. Values for temperature, air flow, agitator speed, pH, dissolved oxygen, ethanol, carbon dioxide, and oxygen compositions of the off-gas, may be automatically recorded, data logged, or recorded manually at sample intervals. Other measurements and observations might be made, e.g. colour, turbidity, gas hold-up, etc., and acid, alkali, and antifoam usage, if pH and foam level are controlled. The liquid sample volume and any preflush volume is recorded and the sample is analysed for yeast dry weight, sugar content, ethanol content, total soluble solids, and possibly inorganic and complex yeast nitrogen. The cells may be observed under the microscope, and their size and morphology noted. Their viability might also be checked by selective staining or plate counting. It is clear that even the simplest of fermentation experiments can generate an enormous amount of data. If the modelling is done well, then the derived information will be useful for the design of batch plants, fed-batch, staged, or continuous stirred-tank processes. For this introductory example only a small sub-set of the data is used. *Figure 1* is a plot of the experimental data. The nomenclature used is given in *Table 1*.

It is usual to start off by making a variety of simplifying assumptions. After an initial look at the results of the first modelling exercise, some of these assumptions are removed and a more complex model evolves. The first assumptions made are:

(i)   the laboratory fermenter is well mixed, that is, conditions are uniform throughout the whole of its interior;

(ii)   yeast cells do not die or become non-viable;

Fermentation modelling

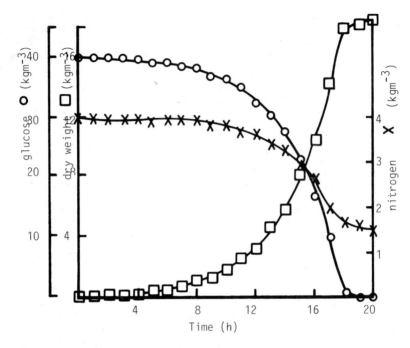

Figure 1. Plot of experimental, data—yeast fermentation.

Table 1. Nomenclature. In the following nomenclature it is always possible to substitute g L^{-1} for kg m^{-3}. In the units given S stands for carbon substrate, N for nitrogen substrate, O for oxygen substrate, P for product, E for enzyme, X_v for viable cells, X_d for non-viable cells and Y for any substance.

a	Area for interphase transport	m^2 m^{-3}
α	Growth associated product formation coefficient	kg P kg X$_v$$^{-1}$
β	Non-growth associated product formation coefficient	kg P kg X$_v$$^{-1}$ h^{-1}
C	Oxygen concentration in the liquid medium	kg O m^{-3}
C^*	Oxygen concentration in the medium when in equilibrium with the gas phase	kg O m^{-3}
C_d	Carbon dioxide concentration in medium	kg CO$_2$ m^{-3}
D	Dilution rate: flow into or through a fermenter divided by the medium volume	h^{-1}
D_c	Critical dilution rate above which washout occurs	h^{-1}
E_o	Total enzyme concentration in a reacting system	g E litre^{-1}
F	Rate of flow of liquid	m^3 h^{-1}
F_i	Bulk liquid flow into a fermenter	m^3 h^{-1}
F_o	Bulk liquid flow out of a fermenter	m^3 h^{-1}
γ	Parameter in expression to ensure product formation stops when substrate is finished	kg S m^{-3}
H	Henry's Law coefficient	kg O m^{-3} bar^{-1}
k	Rate coefficient in Michaelis–Menten equation	g P g E^{-1} s^{-1}
k_d	Coefficient for death rate or rate of conversion of cells to non-viable form	h^{-1}
k_i	Inhibition coefficient	m^3 kg^{-1}
k_L	Mass transfer coefficient for transfer across a phase boundary	m h^{-1}
k_La	Volumetric mass transfer coefficient for transfer across a phase boundary	h^{-1}
k_m	Michaelis–Menten enzyme saturation coefficient	g S litre^{-1}

k_s	Monod cell growth saturation coefficient	kg S m^{-3}
m_o	Maintenance coefficient for cells on oxygen	kg O kgX$_v^{-1}$ h^{-1}
m_s	Maintenace coefficient for cells on carbon substrate	kg S kgX$_v^{-1}$ h^{-1}
N	Concentration of nitrogen substrate in the fermentation medium	kg N m^{-3}
N_e	Number of independent equations in a model	dimensionless
N_{io}	Rate of oxygen transfer from gas to liquid phase	kg O m^{-3} h^{-1}
N_{iy}	Rate of transfer of substance y from gas to liquid phase	kg Y m^{-3} h^{-1}
N_{oc}	Rate of transfer of carbon dioxide from liquid phase to gas phase	kg CO$_2$ m^{-3} h^{-1}
N_{oy}	Rate of transfer of substance y from liquid to gas phase	kg Y m^{-3} h^{-1}
N_v	Number of variables in a mathematical model	dimensionless
$N(0)$	Initial value of nitrogen concentration	kg N m^{-3}
P	Product concentration in the liquid medium	kg P m^{-3}
p_o	Partial pressure of oxygen in the gas phase	bar
$P(0)$	Initial value of product concentration in the fermentation medium	kg P m^{-3}
r_c	Rate of carbon dioxide production	kg CO$_2$ m^{-3} h^{-1}
r_{cy}	Rate of consumption of substance y	kg Y m^{-3} h^{-1}
r_d	Rate of death of viable cells	kg X$_v$ m^{-3} h^{-1}
r_d	Rate of production of non-viable cells	kg X$_d$ m^{-3} h^{-1}
r_{gy}	Rate of generation of substance y	kg Y m^{-3} h^{-1}
r_n	Rate of consumption of nitrogen substrate	kg N m^{-3} h^{-1}
r_o	Rate of consumption of oxygen substrate	kg O m^{-3} h^{-1}
r_p	Rate of product formation	kg P m^{-3} h^{-1}
r_{pc}	Rate of complex energy-requiring product formation	kg P m^{-3} h^{-1}
r_{ps}	Rate of simple energy-producing product formation	kg P m^{-3} h^{-1}
r_s	Rate of carbon substrate consumption	kg S m^{-3} h^{-1}
r_{sm}	Rate of carbon substrate consumption for maintenance	kg S m^{-3} h^{-1}
r_x	Rate of growth (production) of viable cells	kg X$_v$ m^{-3} h^{-1}
S	Carbon substrate concentration in the liquid medium	kg S m^{-3}
S_i	Carbon substrate concentration in the feed to a fermenter	kg S m^{-3}
$S(0)$	Initial value of the carbon substrate concentration in a fermenter	kg S m^{-3}
t	time	h
μ	Specific growth rate of cells	h^{-1}
μ_m	Maximum specific growth rate of cells	h^{-1}
v	Rate (velocity) of an enzyme reaction	g P litre^{-1} s^{-1}
V	Volume of control region	m^3
V_g	Volume of gas	m^3
V_m	Maximum volume of medium in fed-batch fermenter	m^3
V_o	Initial volume of liquid in fed-batch fermenter	m^3
x_d	Concentration of non-viable cells in medium	kg X$_d$ m^{-3}
x_v	Concentration of viable cells in medium	kg X$_v$ m^{-3}
x_{vb}	Concentration of viable cells in medium at end of growth phase and beginning of production phase in fed-batch fermenter	kg X$_v$ m^{-3}
x_{vi}	Concentration of viable cells in feed to a fermenter	kg X$_v$ m^{-3}
$x_v(0)$	Initial value of concentration of viable cells in a fermenter	kg X$_v$ m^{-3}
y	Concentration of a substance y in the medium	kg Y m^{-3}
y_i	Concentration of a substance y in the feed to a fermenter	kg Y m^{-3}
$Y''_{p/o}$	Yield factor for product on oxygen (includes oxygen for energy production for synthesis)	kg P kg O^{-1}
$Y_{p/s}$	Yield coefficient for product on carbon substrates	kg P kg S^{-1}
$Y''_{p/s}$	Yield factor for product on carbon substate (includes carbon substrate to provide energy required for synthesis)	kg P kg S^{-1}
$Y_{pc/s}$	Yield coefficient for complex products on carbon substrate	kg P kg S^{-1}
$Y_{ps/s}$	Yield coefficient for simple breakdown products on carbon substrate	kg P kg S^{-1}
$Y_{x/n}$	Yield coefficient for cells on nitrogen substrate	kg x$_v$ kg N^{-1}
$Y''_{x/o}$	Yield factor for cells on oxygen (including oxygen to provide energy for synthesis)	kg X$_v$ kg O^{-1}
$Y_{x/s}$	Yield coefficient for cells on carbon substrate	kg X$_v$ kg S^{-1}
$Y''_{x/s}$	Yield factor for cells on carbon substrate (includes carbon substrate required to provide energy for synthesis)	kg X$_v$ kg S^{-1}

(iii)   we can ignore production of ethanol;

(iv)   the only two substrates of importance are the carbon source (sugar or molasses) and nitrogen;

(v)   oxygen is provided in excess of the need for metabolism and energy production. (This will only occur for low starting concentrations of carbon substrate.)

This list is not exhaustive, but covers the main assumptions. The next step is to decide what questions need to be answered. At this preliminary stage, typical questions would be:

(i)   What is the maximum growth rate of the yeast?

(ii)   How much of the carbon source is used for growth and how much for maintenance?

(iii)   What is the nitrogen requirement of the yeast?

(iv)   Is this a perfectly normal aerobic growth of yeast with negligible amounts of other products, e.g. ethanol and other alcohols, low-molecular-weight acids, etc. being produced?

We now require a mathematical model. Searching through the textbooks we discover that growth is usually described by a Monod model:

$$r_x = \mu_m S\, x_v/(k_s + S)$$

where r_x is the rate for cell growth (kg cells m^{-3} h^{-1}), S is the substrate concentration (kg carbon source m^{-3}), and x_v is the viable cell concentration (kg cells m^{-3}). The above three are variables which we get from the experimental data; μ_m, the maximum specific growth rate (kg cells kg cells^{-1} h^{-1}), and k_s, the saturation constant for substrate (kg carbon source m^{-3}), are parameters which are characteristic of the yeast and should remain the same if temperature, pH, substrate concentration and other 'environmental' conditions are unaltered. They can then be used for design purposes.

Next we find that substrate consumption rate is given by

$$r_s = r_x/Y'_{x/s} + m_s\, x_v$$

where r_s is the rate of substrate consumption (kg carbon source m^{-3} h^{-1}), $Y'_{x/s}$, is the yield factor for biomass on substrate (kg cells kg carbon source^{-1}), and m_s is the maintenance constant (kg carbon source kg cells^{-1} h^{-1}).
The first is a variable, while the second and third are parameters.
Finally, we discover that for nitrogen consumption

$$r_n = r_x/Y_{x/n}$$

where

r_n = rate of nitrogen consumption (kg nitrogen m^{-3} h^{-1})
$Y_{x/n}$ = yield coefficient for biomass on nitrogen (kg cells kg nitrogen^{-1}).

In this case r_n is the variable while $Y_{x/n}$ is a parameter.

1.1 Units and definitions

The units given above are very detailed; for instance, m_s has the units kg carbon source kg cells^{-1} h^{-1}. In the initial stages of modelling it is good practice to write units in as descriptive a way as possible. Many workers write the units of m_s as kg

$kg^{-1} h^{-1}$, but this obscures the correct definition of m_s as the rate of substrate consumption in kgh^{-1} required to provide energy (by oxidation) for the maintenance of one kg of cells. It might also be tempting to cancel the kg and give m_s the units h^{-1}, which would be wrong.

You will also have noticed that μ_m has the units kg cells kg cells^{-1} h^{-1}. This arises from the definition of any specific rate, which in biological systems is always a rate of something per unit amount of cells. Normally, and quite properly, the two kg units can be cancelled since they are both kg of cells.

$Y'_{x/s}$ is a yield factor and is so indicated by the prime superscript. It is the ratio of mass of cells formed to the mass of carbon substrate required, both to be incorporated in the cell matter by normal biochemical reactions and to be oxidized to CO_2 and water (or ethanol, lactic acid, etc.,) to provide the energy necessary to synthesize the cells. Note that it is not a stoichiometric coefficient.

$Y_{x/n}$ This is true stoichiometric coefficient, which is why it is called the yield coefficient. It can be obtained directly from the biochemical equation for synthesis of cells from the nitrogen source. For example,

$$CH_2O + 0.2NH_4 + e- = CH_{1.8}O_{0.5}N_{0.2} + 0.5H_2O$$
$$(30) \qquad 0.2 \times (18) \qquad (24.6) \qquad 0.5 \times (18)$$

represents a possible synthesis of yeast (molecular weights are given in brackets).
The yield coefficients for cells on nitrogen are:
in terms of elemental nitrogen $\quad 24.6/(0.2 \times 14) = 8.8$
in terms of ammonium ion $\quad\quad 24.6/(0.2 \times 18) = 6.8$
in terms of anhydrous ammonia $\quad 24.6/(0.2 \times 17) = 7.2$

1.2 A little calculus

There are five variables in the equations we have used so far. S and x_v we already have from analyses of concentration and dry weight, but how do we find the rates of growth and substrate consumption r_x, r_s and r_n from the experimental data? To do this we need to make use of a little more mathematics, the so-called balance equation. If we have a closed system like our batch fermenter then we can say that:

Rate of accumulation or   Rate of growth or
depletion in the system   =   consumpton

The rate of accumulation or depletion is the change in concentration divided by the time interval over which it occurs, i.e.:

$$\frac{Concentration_2 - Concentration_1}{Time_2 - Time_1}$$

or for cell dry weight:

$$\frac{x_{v2} - x_{v1}}{t_2 - t_1}$$

where x_{v1}, x_{v2} are the biomass concentration at t_1 and t_2 respectively (kg m^{-3}), and t_1, t_2 are time (in hours) at which samples 1 and 2 were taken. This can be abbreviated to $\Delta x_v/\Delta t$ or, in the limit, as t becomes infinitesimally small, dx_v/dt.

69

→ Our model equations therefore become:

for yeast:
$$\frac{dx_v}{dt} = r_x = \mu_m \frac{S\,x_v}{(k_s + S)} \tag{1}$$

for carbon substrate:
$$\frac{dS}{dt} = -r_s = -\left\{ \frac{r_x}{Y'_{x/s}} + m_s\,x_v \right\} \tag{2}$$

and
for nitrogen substrate:
$$\frac{dN}{dt} = -r_n = \frac{-r_x}{Y_{x/n}} \tag{3}$$

Note the minus sign for depletion or consumption (concentration falling). The three equations given above represent the basic model for the fermentation and, providing that:

(i)   we can determine values of the parameters μ_m, k_s, $Y'_{x/s}$, m_s and $Y_{x/n}$; and

(ii)   the model is sufficiently 'robust', i.e., it will represent what happens in the experimental situation for a variety of conditions;

we can use it as an adequate representation of the fermentation.

   The data we have is organized as sets of data points, each set corresponding to one particular value of elapsed time. The full set of data points and the units are:

time t (hours or minutes)
cell concentration x_v (g litre^{-1} or kg m^{-3})
carbon substrate S (g litre^{-1} or kg m^{-3})
concentration
nitrogen substrate N (g litre^{-1} or kg m^{-3})
concentration

   The values of the derivatives of x_v, S, and N, i.e., dx_v/dt, dS/dt and dN/dt which are (but only for a closed-batch fermentation) equal to the three rates r_x, r_s, and r_n are as defined previously. As shown in *Figure 2* for x_v, we can determine the value of the derivative by various methods of differing degrees of accuracy and convenience. Unless the data points lie on a smooth curve, which is most unlikely, then a smooth curve should be drawn through the experimental points as has been done in *Figure 1* and is done in *Figure 2*.

   The two simplest methods are:

(i)   to hand-draw a tangent to the smooth curve drawn through the points and measure its slope.

(ii)   to take the difference between the values equally spaced on either side of the data point being analysed and divide by the appropriate time interval, which in this case is 2 h: (numerical differentiation).

A third, more complex method is to fit a mathematical equation to three or more data points around the point being analysed and then, mathematically, compute the required slope (tangent). There are various methods of doing this and the equation may be purely empirical e.g., a quadratic, or cubic or higher-order polynomial, or it may be a mathematical form more related to the phenomena underlying the data, e.g. in this case exponential growth. The three methods are illustrated in *Figure 2* and the results are, using the best-fit curve through the data points:

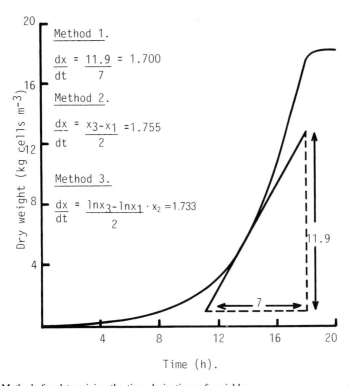

Figure 2. Methods for determining the time derivatives of variables.

Table 2. Calculation of derivatives using numerical differentiation.

Time (h)	x_v (kg cells m^{-3})	$r_x = dx_v/dt$ (kg cells $m^{-3} h^{-1}$)	$\mu = r_x/x_v$ (h^{-1})	$1/\mu$ (h)	S (kg glucose m^{-3})	$1/S$ (m^3 kg glucose^{-1})
0	0.100	–	–	–	40.00	0.0250
1	0.134	0.040	0.298	3.356	39.93	0.0250
2	0.180	0.054	0.300	3.333	39.83	0.0251
3	0.241	0.072	0.299	3.345	39.70	0.0252
4	0.323	0.096	0.297	3.367	39.50	0.0253
5	0.433	0.129	0.298	3.356	39.30	0.0254
6	0.581	0.172	0.296	3.378	38.97	0.0257
7	0.778	0.230	0.296	3.378	38.50	0.0260
8	1.040	0.311	0.299	3.344	38.00	0.0263
9	1.400	0.415	0.296	3.378	37.20	0.0269
10	1.870	0.550	0.294	3.401	36.20	0.0276
11	2.500	0.740	0.296	3.378	34.80	0.0287
12	3.350	0.995	0.297	3.367	32.90	0.0304
13	4.490	1.325	0.295	3.390	30.50	0.0328
14	6.000	1.755	0.293	3.413	27.20	0.0368
15	8.000	2.350	0.294	3.401	22.80	0.0438
16	10.70	3.050	0.285	3.509	17.10	0.0585
17	14.10	3.600	0.255	3.922	9.60	0.104
18	17.90	2.100	0.117	8.547	1.11	0.901

Table 3. Comparison of method for obtaining derivatives. Values of $\frac{dx}{dt}$ (kg cells m^{-3} h^{-1})

Time (h)	Method 1	Method 2	Method 3
7	0.266	0.230	0.255
10	0.542	0.550	0.515
12	0.981	0.995	0.890
16	3.032	3.050	3.070
18	2.333	2.1000	1.900

(i) *Hand-drawn tangent.*

$$\frac{dx_v}{dt} = \frac{12.9 - 1.0}{7} = 1.7 \text{ kg m}^{-3} \text{ h}^{-1}$$

(ii) *Numerical differentiation.*

$$\frac{dx_v}{dt} = \frac{8.0 - 4.49}{2} = 1.755 \text{ kg m}^{-3} \text{ h}^{-1}$$

(iii) *Curve fitting.* An appropriate theoretical curve might be that for exponential growth, $x_v = a \exp(bt)$, where a and b are constants, which when differentiated gives

$$\frac{dx_v}{dt} = ab \exp(bt) = bx_v$$

so that to find $dx_v dt$ we need the value of b.

By taking logarithms of either side of the theoretical equation, $\ln(x) = \ln(a) + bt$, hence $b(t_3 - t_1) = \ln(x_3) - \ln(x_1)$, ($\ln(a)$ cancels out) so that

$$b = \frac{2.079 - 1.502}{2} = 0.289 \text{ h}^{-1}$$

and

$$\frac{dx_v}{dt} = 0.289 \times 6.0 = 1.733 \text{ kg m}^{-3} \text{ h}^{-1}$$

(Note that the closer the true form of the experimental curve is to the selected curve the more accurate will be the estimate.)

The first method is very easy, but accuracy is low. The second works well enough for limited growth but can give quite inaccurate results when exponential growth is occurring or when the data points are spaced widely apart. The third method is the most accurate and is normally used in automatic computational methods. The second method is a variant of the third in which the curve fitted is a straight line. In this section we have used the second method to determine all the derivatives. The results of these calculations are shown in the first three columns of *Table 2*. To indicate how the different methods compare, *Table 3* shows the results from all the methods for selected representative points. If you wish to use the experimental data points to find the values of the derivatives, then only the third method of fitting a mathematical equation can be used.

1.3 Estimation of parameters

To find the parameters we make use of the processed data points listed in *Table 2* and the mathematical model (equations 1, 2, and 3).

1.3.1 *Estimation of* μ_m *and* k_s

$$r_x = \frac{dx_v}{dt} = \frac{\mu_m \, S \, x_v}{k_s + S} \tag{1}$$

There are three variables in this equation, all of whose values change from data point to data point. Since we have the products of variables, e.g. $S \, x_v$, and reciprocals, e.g. $x_v/(k_s + S)$, it cannot be handled straightforwardly. The trick is to rearrange the variables in groups or change them in some way so that the equation looks like something we are used to handling. The simplest equation to work with is one which when plotted is a straight line ($y = m \, x + c$, where y and x are the variables, m is the slope, and c the intercept). This process is called transforming the equation, and the new groupings of or changes in variables are called transformed variables. The first obvious grouping is to divide both sides of equation (1) by x_v to give

$$\frac{r_x}{x_v} = \frac{\mu_m \, S}{k_s + S} \tag{1a}$$

Notice how we have already simplified the problem by reducing the number of variables from three untransformed to two transformed, i.e. r_x/x_v and S. When we group variables together, as we did by dividing r_x by x_v, the result is often given a new symbol. In this case the conventional symbol is μ, so that

$$\mu = \frac{r_x}{x_v} = \frac{dx_v/dt}{x_v}$$

(for a closed batch process only). Values of μ are listed in *Table 2*.

Remember that μ is a variable even though it is called the specific growth rate since it depends on the value of S. We could now plot μ against S and then find values of the parameters μ_m and k_s which give the theoretical curve which best fits the data. We have done this in *Figure 3*; it is obviously a messy business. It is better to look for another transformation which gives a straight line. If we take the reciprocal of both sides of equation (1a), we get

$$\frac{x_v}{r_x} = \frac{1}{\mu} = \frac{k_s + S}{\mu_m S}$$

i.e., $\tag{1b}$

$$\frac{1}{\mu} = \frac{k_s}{\mu_m S} \frac{1}{S} + \frac{1}{\mu_m}$$

and this is our desired goal. Equation (1b) states that if our model (equation 1) holds up then when we plot

$$\frac{1}{\mu} \left\{ = \frac{x_v}{dx_v/dt} \text{ for a closed batch} \right\}$$

against $1/S$ on arithmetic graph paper then we should get a straight line with slope k_s/μ_m and intercept on the $1/\mu$ axis of $\frac{1}{\mu_m}$. Hence we can easily find k_s and μ_m. This plot, called a Lineweaver–Burke plot after the people who first suggested it, is shown in *Figure 4*. The values obtained from the plot are:
Intercept (= $1/\mu_m$ = 3.4 h, slope (= k_s/μ_m) = 5.67 kg h^{-1} m^{-3}. Hence μ_m = 0.294 h^{-1} and k_s = 1.67 kg m^{-3}. You will notice that because of the geometry of the plot a value for $1/k_s$ can be read directly off the abscissa (horizontal axis).

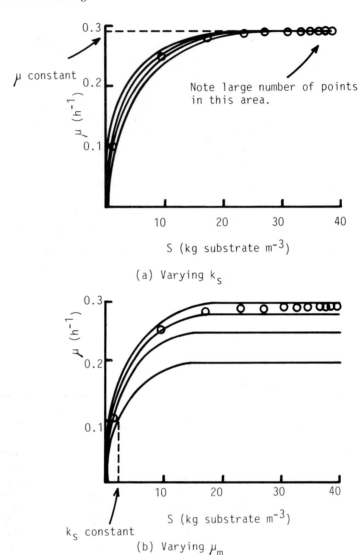

Note large number of points in this area.

(a) Varying k_s

(b) Varying μ_m

Figure 3. Determination of μ_m and k_s by curve fitting.

1.3.2 *Estimation of $Y'_{x/s}$, m_s, $Y_{x/n}$*

The other parameters are found in a similar way. *Table 4* lists values for all the variables r_x, r_s, r_n, x_v, S, N, t, using the second method for finding the derivatives dx_v/dt, dS/dt and dN/dt.

Since

$$r_s = \frac{r_x}{Y'_{x/s}} + m_s/x_v \tag{2}$$

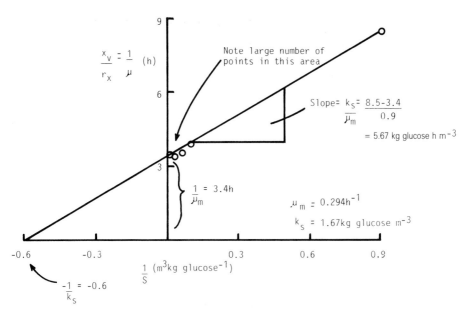

Figure 4. Determination of μ_m and k_s by Lineweaver–Burke plot.

Table 4. Fully processed data for yeast fermentation.

t (h)	x_v (kg m^{-3})	S (kg m^{-3})	N (kg m^{-3})	r_x (kg m^{-3} h^{-1})	r_s (kg m^{-3} h^{-1})	r_N (kg m^{-3} h^{-1})
0	0.100	40.00	4.00			
1	0.134	39.93	4.00	0.040	0.085	0.005
2	0.180	39.83	3.99	0.054	0.115	0.010
3	0.241	39.70	3.98	0.072	0.165	0.010
4	0.373	39.50	3.97	0.096	0.200	0.010
5	0.433	39.30	3.96	0.129	0.265	0.015
6	0.581	38.97	3.94	0.172	0.400	0.020
7	0.778	38.50	3.92	0.230	0.485	0.030
8	1.040	38.00	3.88	0.311	0.650	0.040
9	1.400	37.20	3.84	0.415	0.900	0.050
10	1.870	36.20	3.78	0.550	1.200	0.070
11	2.500	34.80	3.70	0.740	1.650	0.095
12	3.350	32.90	3.59	0.995	2.150	0.130
13	4.490	30.50	3.44	1.325	2.850	0.175
14	6.000	27.20	3.24	1.755	3.850	0.235
15	8.000	22.80	2.97	2.350	5.050	0.335
16	10.700	17.10	2.57	3.050	6.600	0.440
17	14.100	9.60	2.09	3.600	7.995	0.535
18	17.900	1.11	1.50	2.100	4.800	0.300
19	18.300	0.00	1.49			
20	18.300	0.00	1.48			

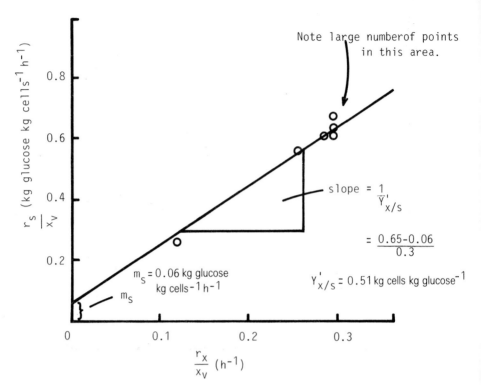

Figure 5. Straight line plot for determining $Y'_{x/s}$ and m_s.

We have three variables and have to transform them to get a simple, linear relationship between two transformed variables. The most obvious is to divide through by x_v to give

$$\frac{r_s}{x_v} = \frac{1}{Y'_{x/s}} \cdot \frac{r_x}{x_v} + m_s \tag{2a}$$

The transformed variables are r_x/x_v and r_s/x_v. By plotting these two transformed variables as shown in *Figure 5*, we obtain the slope of the straight line, $1/Y'_{x/s} = 1.96$ kg carbon source kg cells^{-1}, so that the yield factor, $Y'_{x/s}$, $= 0.51$ kg cells kg carbon source^{-1} and the intercept, $m_s = 0.06$ kg carbon source kg cells^{-1} h^{-1}.

Similarly, using the nitrogen balance we have

$$\frac{dN}{dt} = r_n = \frac{r_x}{Y_{x/n}} \tag{3}$$

so that a plot of r_n versus r_x, shown in *Figure 6*, gives a straight line through the points. According to our model this should go through the origin, but as can be seen, the points fall on a slight curve. This curvative indicates that the model for nitrogen consumption in the batch system is not an ideal representation. The nature of the deviation from the straight line indicates a loss of nitrogen from the system which is quantitatively larger at higher growth rates. This could be due to ammonia loss in the exit gas stream, or possibly some experimental losses during the nitrogen assay. Assuming that the straight-line relationship holds and we force it to go through the origin, we get a value of yield coefficient $Y_{x/n}$ equal to 6.84, which is slightly less than we might expect from theory.

76

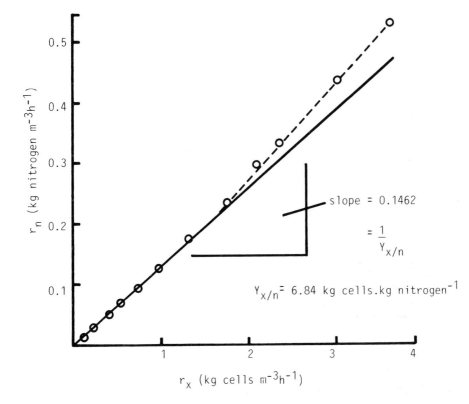

Figure 6. Straight line plot for determining $Y'_{x/n}$.

1.4 Model of experimental yeast fermentation

The final result of all these manipulations is the model which describes the fermentation which is given below:

Balance equations

Yeast cells $\qquad \dfrac{dx_v}{dt} = r_x$

Carbon substrate $\qquad \dfrac{dS}{dt} = -r_s$

Nitrogen substrate $\qquad \dfrac{dN}{dt} = -r_n$

Rate equations

$$r_s = \frac{0.294S}{1.67 + S}\ x_v$$

$$r_s = \frac{r_x}{0.51} + 0.06x$$

$$r_n = \frac{r_x}{6.84}$$

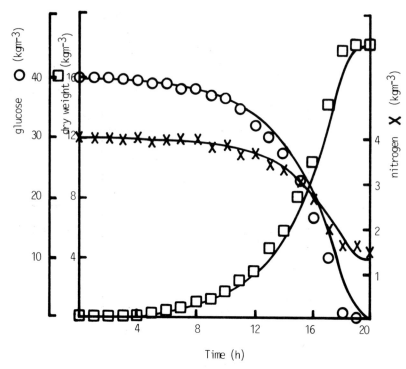

Figure 7. Comparison of model with experimental results.

Initial conditions

$x_v(0)$ (time zero) = 0.10
$S(0)$ (time zero) = 40.00
$N(0)$ (time zero) = 4.00

Constraints

$x > 0$
$0 \leq S \leq S(0)$
$0 \leq N \leq N(0)$

   The predictions of the model can now be compared with the experimental data as shown in *Figure 7*. It is not surprising that the model gives good predictions for this data, since the values of the parameters were obtained using the same data points. The value of the model will depend on how well it predicts other systems with different conditions, e.g. different starting values (of $x(0)$, $S(0)$, $N(0)$) or different fermenter configurations (e.g. fed-batch, staged batch, continuous single vessel, or continuous multiple vessel). It must also be realized that the values of the parameters particularly μ_m, k_s and m_s will depend on the temperature, pH and other environmental conditions.

2. BASIC METHODOLOGY FOR WRITING MODEL EQUATIONS

The procedure for writing model equations has been briefly touched on in the introduction. In this section, the formal method is given as a set of rules.

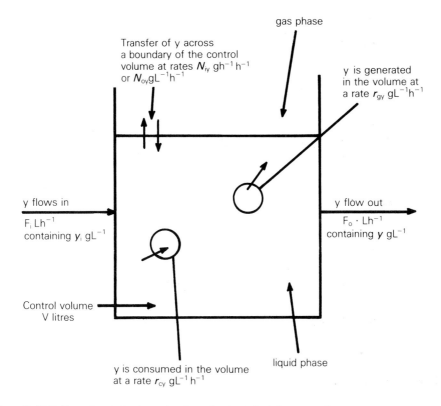

Figure 8. Definition of control volume, inputs, and outputs for balance equations.

2.1 Balance equations

The balance equation is an accounting relationship between the rates at which a quantity y (y could be cells, carbon substrate, penicillin, etc.) accumulates in a control volume, and rates at which this same quantity flows into and out of the volume with a bulk liquid flow, is generated or consumed within the region or is transferred by mass transfer or diffusion across a surface defining the volume. *Figure 8* gives the general idea. The main points to remember about this are:

(i)   The control volume must be defined precisely, usually in terms of phase boundaries (gas/liquid, solid/liquid or rarely liquid/liquid).

(ii)   The control volume can change with time (if F_i differs from F_o).

(iii)   The concentration of y must be the same everywhere throughout the volume. (This is usually referred to as a well-mixed volume.) This implies that the concentration of y within the volume is the same as the concentration leaving.

(iv)   Rates of generation, consumption, or transfer across a boundary other than by bulk flow are always given in terms of unit volume, so to find the total generated, consumed, or transferred, the amount per unit volume must be multiplied by the total volume.

We can immediately write the balance as:

$$\text{Rate of accumulation} = \text{rate in} - \text{rate out}$$

Table 5. Symbols used in reaction equations

Volume of liquid	V	m^{-3}
Volume of gas	V_g	m^{-3}
Time	t	h
Viable cell concentration	x_v	$kg\ m^{-3}$
Non-viable (dead) cell concentration	x_d	$kg\ m^{-3}$
Carbon substrate concentration	S	$kg\ m^{-3}$
Nitrogen substrate concentration	N	$kg\ m^{-3}$
Dissolved oxygen concentration	C	$kg\ m^{-3}$
Dissolved carbon dioxide concentration	C_d	$kg\ m^{-3}\ h^{-1}$
Rate of cell growth (generation)	r_x	$kg\ m^{-3}\ h^{-1}$
Rate of cell death (consumption)	r_d	$kg\ m^{-3}\ h^{-1}$
Rate of substrate uptake (consumption)	r_s	$kg\ m^{-3}\ h^{-1}$
Rate of oxygen uptake (consumption)	r_o	$kg\ m^{-3}\ h^{-1}$
Rate of CO_2 production (generation)	r_c	$kg\ m^{-3}\ h^{-1}$
Rate of product formation (generation)	r_p	$kg\ m^{-3}\ h^{-1}$
Rate of oxygen transfer (gas to liquid)	N_{io}	$kg\ m^{-3}\ h^{-1}$
Rate of CO_2 transfer (liquid to gas)	N_{oc}	$kg\ m^{-3}\ h^{-1}$

Table 6. Typical balance equations

Viable cells (no cells entering)

$$\frac{d(Vx_v)}{dt} = 0 + V\,r_x - F_o\,x_v - V\,r_d$$

Non-viable cells (no cell entering)

$$\frac{d(Vx_d)}{dt} = 0 + V\,r_d - F_o\,x_d$$

Carbon substrate

$$\frac{d(VS)}{dt} = F_i\,S_i - V\,r_s - F_oS$$

Product

$$\frac{d(VP)}{dt} = F_i\,P_i + V\,r_p - F_o\,P$$

Oxygen

$$\frac{d(VC)}{dt} = F_i\,C_i + V\,N_{io} - V\,r_o - F_o\,C$$

Carbon dioxide

$$\frac{d(VC_d)}{dt} = F_i\,C_{di} + V\,r_c - V\,N_{oc} - F_o\,C_d$$

Rate in =   bulk flow into the volume
            + generation within the volume
            + transfer into the volume across the boundaries other than by bulk flow

Rate out =   bulk flow out
            + consumption within the volume
            + transfer out of the volume across the boundaries other than by bulk flow

Using symbols is a kind of shorthand but, of course, until you are familiar with the symbols more thought is needed.

The rate of accumulation in the volume is the amount in the volume at time t_2, say V_2y_2, minus the amount at time t_1, V_1y_1 divided by the time interval $t_2 - t_1$, i.e.

$$\text{rate of accumulation} = \frac{V_2y_2 - V_1y_1}{t_2 - t_1}$$

(Verify that this has the units kg (of y) h^{-1}.)

In order to get a picture of what is happening at an instant in time, we shrink the time interval down to an infinitesimal amount, dt, and the amount shrinks to an equivalent infinitesimal amount $d(Vy)$, so that the rate of accumulation is $d(Vy)/dt$.

We can now write the balance equation for the quantity y using the nomenclature of *Figure 8* as:

$$\frac{d(Vy)}{dt} = (F_i\,y_i + V\,r_{gy} + V\,N_{iy}) - (F_o\,y + V\,r_{cy} + V\,N_{oy})$$

(Verify that the units of every term separated by a $+$, $-$, or $=$ sign are the same). Note that y_i is the concentration of the appropriate quantity entering the fermenter, y is the concentration of the same quantity both in the fermenter and in the stream leaving the fermenter.

What we have written is a linear ordinary differential equation for a quantity, y which can be expressed in mass or energy units (usually grams or joules). (We only deal with mass units in this chapter.) y can stand for anything which takes part in the biochemical reactions in a fermenter. The symbols are fairly well standardized and are usually as listed in *Table 5*.

Some typical balance equations are given in *Table 6*.

Make sure that you understand the significance of each group of terms in the above equations, and satisfy yourself that the units of each group (separated by $+$, $-$, or $=$) are the same within any one balance equation (using the full unit definitions e.g. kg viable cell m^{-3}, not kg m^{-3}. One other balance equation which is important is the volume balance which is simply (*Figure 8*).

$$\frac{dV}{dt} = F_i - F_o$$

2.2 Rate equations

In the conservation equations, the rate of accumulation terms and the bulk flow in and out terms are written in terms of measurable quantities such as volumes, flow rates and concentrations. The rate terms, i.e. rates of generation and consumption and rates of mass transfer across the volume boundaries, are not directly measurable and have to be expressed in terms of known or measurable variables. The rates of generation or consumption are known kinetic rates, whilst the others are mass transfer rates. Most model making and the research which underpins it are concerned with determining the forms of these expressions. Two examples will serve to illustrate the point.

(i) *Rate of cell growth (generation)*. A common expression is the Monod equation

$$r_x = \frac{\mu_m S}{k_s + S}\,x_v$$

which relates the rate, r_x, to two variables we can measure, carbon substrate concentration, S, and cell concentration, x_v. The expression also contains two parameters, μ_m and k_s, which are assumed to be constants for a particular microorganism at constant environmental conditions, temperature, pH, etc.

(ii) *Rate of oxygen transfer from gas to liquid.* The common expression for this rate of mass transfer is

$$N_{io} = k_L a \, (C^* - C)$$

which again relates the rate, N_{io}, to two variables, C, the oxygen concentration in the bulk liquid and C^*, the oxygen concentration in the liquid at the interface between the gas and liquid. C can be measured. C^* cannot be measured but it is a variable and we will consider it in the next section. The two parameters are k_L, the mass transfer coefficent and a, the interface area per unit volume of liquid. They are usually regarded as a single quantity, $k_L a$, called the volumetric mass transfer coefficient. $k_L a$ depends upon aeration rates, power input and the physical properties such as viscosity, density, etc. of the medium. Various correlations exist for estimating $k_L a$.

2.3 Thermodynamic equations

Sometimes, as in the expression for N_{io} in the previous section, the rate expression or the balance equation contain a variable which is not directly or not conveniently measured. In such a case we look for a thermodynamic expression which relates the value of the variable to the value of a known or measurable variable. This relationship will be an equilibrum relationship, hence the term thermodynamic. The thermodynamic equation used for C^* is Henry's Law which relates the concentration of a sparingly soluble gas in a liquid to its partial pressure (concentration) in the gas phase which is in equilibrium with the liquid. Thus

$$C^* = H \, p_0$$

Where H is the Henry's law constant and p_0 is the partial pressure of oxygen in the gas phase. p_0 in turn can be related to the oxygen concentration in the gas. It should be noted that these thermodynamic equations imply equilibrium between the variables e.g., variation in the partial pressure of oxygen in the gas phase cause an instantaneous corresponding variation in concentration at the liquid interface.

2.4 Checking the equations

There are two main preliminary checks that can be made for dimensional consistency and degrees of freedom.

2.4.1 *Dimensional consistency*

When writing equations, the dimensions of every symbol used should be noted in a table of nomenclature which is built up as the equations are written. This was done in Section 2.1. When using the dimension gram (g), litre or m^3, state what substance you are referring to e.g., g carbon substrate, m^3 medium, cm^3 cell volume, etc. Remember that in the accumulation term dy/dt, dy has the dimensions of y and dt of t (i.e., hours). In any equation, terms separated by $+$, $-$, or $=$ should have the same dimensions. Groups of terms of which the logarithm is taken must be dimensionless, and any group of terms which is an exponent on a quantity (raised to the power of that group of terms or the exponential function of the group) must also be dimensionless.

Two examples are given below.

Example 1.

$$\frac{1}{\mu} = \frac{k_s}{\mu_m} \frac{1}{S} + \frac{1}{\mu_m}$$

$k_s/\mu_m S$ must have the same dimensions as $1/\mu$ and $1/\mu_m$, which means k_s and S must have the same dimensions (so they cancel out). This is also obvious from the definition of $\mu = \mu_m S/(k_s + S)$. Since k_s and S are separated by $+$ they have the same dimensions (g substrate), (L medium)$^{-1}$ or (kg substrate) (m^{-3} medium).

Example 2. The solution of the equation for exponential cell growth in batch culture,

$$\frac{dx_v}{dt} = \mu_m x_v$$

is

$$\ln(x_v/x_v(0)) = \mu_m t$$

or

$$x_v = x_v(0) \exp(\mu_m t) = x_v(0) e^{\mu_m t}$$

note that $x_v/x_v(0)$ must be dimensionless since it is the argument of the log function and $\mu_m t$ must also be dimensionless (h^{-1}h) since it is an exponent (power) or the argument of the exponential function.

2.4.2 Degrees of freedom

If we count the number of equations in our model, N_e, and the number of variables, N_v, then the difference $N_v - N_e$ is the number of degrees of freedom which is the number of variables that we can independently fix. For a batch fermentation there should be no degrees of freedom. This means that once the simulation of batch has started by setting the starting conditions (initial conditions) and solving the equations, the simulation proceeds until the batch ends (by running up against a constraint). With other fermentations the number of degrees of freedom should be the number of variables that a person running the plant can independently fix. In a single-stage continuous fermentation without re-cycle, for example, an operator could fix the inlet and outlet flows and the concentrations of all variables in the inlet flow. This number would be the number of degrees of freedom.

Example 1. We use the model of the batch yeast fermentation discussed in Section 1. If we take the first three equations

$$\frac{dx_v}{dt} = r_x, \quad \frac{dS}{dt} = -r_s, \quad \frac{dN}{dt} = -r_n$$

then we have three equations and six variables, x_v, S, N, r_x, r_s, r_n, so we have three degrees of freedom. We cannot independently fix any of these, so we are obviously short of three equations to complete the model. These are the three rate equations

$$r_x = \frac{0.0294\ S}{1.67 + S}\ x_v, \qquad r_s = \frac{r_x}{Y'_{x/s}} + 0.06\ x_v$$

and

$$r_n = \frac{r_x}{Y'_{x/n}}$$

We have not added to the number of variables, so we now have six equations, six variables and no degrees of freedom. The model can be solved.

Example 2. The equations for a generalized fermenter are given in *Table 6.* If we assume that we are interested only in the cells and carbon substrate and that cell death does not occur, that there is no product formation, and that oxygen is always in excess, so that the medium is fully saturated (i.e. $C = C^*$), then we can describe this situation using three balance equations:

$$\frac{d(Vx_v)}{dt} = Vr_x - F_o x_v$$

$$\frac{d(VS)}{dt} = F_i S_i - Vr_s - F_o S$$

$$\frac{dV}{dt} = F_i - F_o$$

We have three equations and eight variables x_v, S, V, F_i, F_o, S_i, r_x and r_s, so that we have five degrees of freedom. This means we have to be able to independently fix five of the variables. This is possible for F_i, F_o and S_i, (but note that if we do this we cannot independently vary V since this is fixed by the volume balance). We could alternatively manipulate F_i, V and S_i, but then F_o would be a dependent variable. We are obviously two equations short. Again it is the rate equations we are missing, namely:

$$r_x = \frac{\mu_m\ S}{k_s + S}\ x_v, \quad r_s = \frac{r_x}{Y'_{x/s}} + m_s\ x_v$$

which gives us five equations, eight variables and the required three degrees of freedom.

3. RATE EQUATIONS

As was indicated in Section 2.2, determining the form of the rate equations is an important part of fermentation research. Fortunately, we can draw on work which has been carried out for over a century and in this section we consider just a few of the more common forms of the rate equations.

3.1 Cell growth

The Michaelis–Menten equation for single substrate enzyme kinetics,

$$v = \frac{k\ E_o\ S}{k_m + S}$$

can be found in any elementary biochemistry text book. In this,

v	is the velocity of the reaction	g product $L^{-1} s^{-1}$
k	is the reaction rate constant	g product (g Enz.)$^{-1} s^{-1}$
E_o	is the total enzyme amount	g enzyme L^{-1}
S	is the substrate concentration	g L^{-1}
k_m	is the saturation constant	g L^{-1}

Note that $k\ E_o$ is the maximum velocity. Since most biochemical reactions are enzyme catalysed it was reasonable to expect that where one of the substrates in a medium eventually limits the growth of a microorganism, the expression for growth rate of the microorganism would be related to this limiting substrate concentration. Furthermore, the amount of enzymes for growth can be expected to be proportional to the amount of microorganisms present. Thus the generally accepted expression for cell growth with a single limiting substrate is:

$$r_x = \frac{\mu_m S}{k_s + S} x_v$$

Here x_v is the cell concentration		kg cells m^{-3}
S is the limiting substrate concentration		kg substrate m^{-3}
k_s is the saturation constant		kg substrate m^{-3}
and μ_m is the maximum specific growth rate		h^{-1}

The principal characteristics of this expression are shown in *Figure 9*. Note particularly the effect of k_s on growth rate. The smaller k_s, the more effective the cell is at growing at high rate at a low concentration of the limiting substrate. In a natural environment

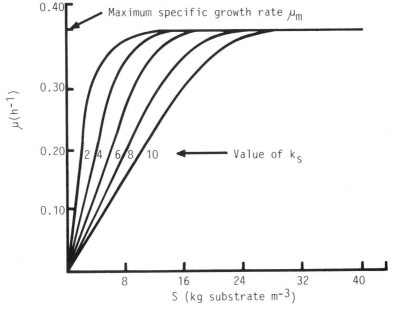

Figure 9. Plot of Monod equation for cell growth.

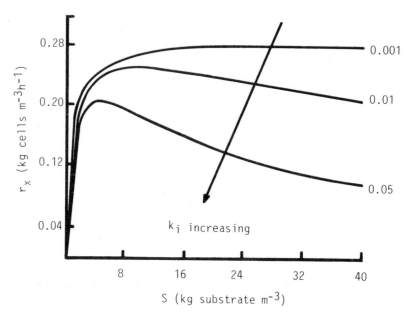

Figure 10. Cell growth rates for substrate inhibition.

microorganisms tend to evolve with a k_s value two orders of magnitude (a factor of 100) less than the normal concentration of their limiting substrate. Other points to notice are:

(i) k_s is the value of the limiting substrate concentration at which the growth rate is half the maximum.

(ii)   the rate of growth is directly proportional to the cell concentration.

Other expressions with similar properties to Monod's have been used to model cell growth, but are rarely used.

3.1.1 *Substrate inhibition*

Monod's equation implies that the cells grow at their maximum rate however high the substrate concentration. But, high concentrations of sugar, salt, acetic acid, etc., can stop cell growth. This is known as substrate inhibition and is modelled by a Monod-type equation with an S^2 term in the denominator, viz.

$$r_x = \mu_m \frac{S}{k_s + S + k_i S^2} x_v$$

The properties of this expression are shown in *Figure 10* for a fixed value of k_s and x_v and varying inhibition constant k_i.

3.1.2 *Product inhibition*

Many metabolic products of cell growth, such as ethanol, lactic acid, etc., inhibit cell growth. The effect of this type of inhibition is usually modelled by multiplying the cell growth expression by a factor which depends on the inhibitory product concentration.

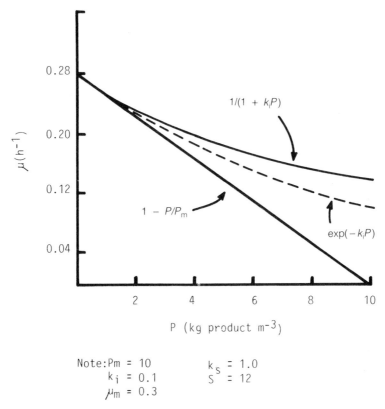

Figure 11. Product inhibition factors.

Note: Pm = 10 k_s = 1.0
      k_i = 0.1   S = 12
      μ_m = 0.3

If P is the inhibitory product concentration then various forms of this factor are:

$$1 - P/P_m, \quad 1/(1 + k_iP), \quad \exp(-k_iP)$$

The effect of product concentration on cell growth rate for a fixed cell and substrate concentration is shown in *Figure 11* for the three factors.

3.2 Substrate utilization: carbon substrate

Carbon is the major constituent by weight of most microorganisms and biological products. It is also the major source of energy for driving the cell processes.

As shown in *Figure 12*, carbon substrate is consumed for the following purposes:

(i)	to synthesise new cell material at a rate	$r_x/Y_{x/s}$
(ii)	to provide energy for the synthesis of new cell material at a rate	$a_x\, r_x$
(iii)	to synthesize excreted complex biochemicals at a rate	$r_{pc}/Y_{pc/s}$
(iv)	to provide energy for the synthesis of excreted complex biochemicals at a rate	$a_p\, r_{pc}$
(v)	to provide energy for cell maintenance at a rate	r_{sm}

Note that $Y_{x/s}$ and $Y_{pc/s}$ are true stoichiometric coefficients, kg (kg substrate)$^{-1}$, and

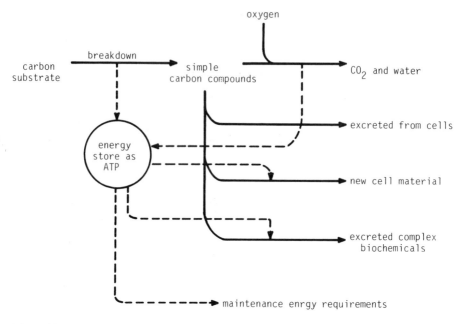

Figure 12. Uses of carbon substrate in the cell.

that we distinguish between complex products, rate r_{pc} and yield coefficient $Y_{pc/s}$, and simple breakdown products, r_{ps} and $Y_{ps/s}$, (see later).

Hence we can state

$$r_s = \left(\frac{r_x}{Y_{x/s}} + a_x r_x \right) + \left(\frac{r_{pc}}{Y_{pc/s}} + a_p r_{pc} \right) + r_{sm}$$

where r_{pc} refers to energy-consuming complex products. Since a_x and a_p are not generally available in the literature (their dimensions are kg substrate for energy kg^{-1}) they are normally combined with the yield coefficient to give a yield factor so that

$$\frac{1}{Y'_{x/s}} = \frac{1}{Y_{x/s}} + a_x$$

and

$$\frac{1}{Y'_{p/s}} = \frac{1}{Y_{pc/s}} + a_p$$

where $Y'_{x/s}$ and $Y'_{p/s}$ are yield factors, not yield coefficients. Thus

$$r_s = \frac{r_x}{Y'_{x/s}} + \frac{r_p}{Y'_{p/s}} + r_{sm}$$

Note that we do not need to make a distinction regarding complex products and simple breakdown products when are are dealing with yield factors denoted by a prime. Generally r_{sm} is assumed to be directly proportional to the amount of cells and is given by

$$r_{sm} = m_s x_v$$

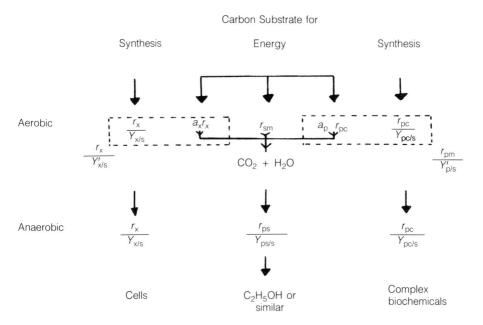

Figure 13. Substrate consumption for aerobic and anaerobic oxidation.

m_s is the maintenance coefficient and has the units kg substrate (kg cell)$^{-1}$h^{-1}.

When we are considering anaerobic operation, we do not need to consider the consumption of energy for maintenance or synthesis of cells and biochemicals since all the energy is accounted for by considering the rate of production, r_p, of the simple carbon compounds, In these circumstances, substrate is consumed for the following purposes:

(i) to synthesize new cell material at a rate $\qquad$ $r_x/Y_{x/s}$
(ii) to synthesize excreted complex biochemicals at a rate $\qquad$ $r_{pc}/Y_{pc/s}$
(iii) to provide all the energy requirements of the cell at a rate $\qquad$ $r_{ps}/Y_{ps/s}$

so that the total rate of substrate consumption is directly related to the cell growth and product formation by true stoichiometric yield coefficients. The equation is therefore

$$r_s = \frac{r_x}{Y_{x/s}} + \frac{r_{pc}}{Y_{pc/s}} + \frac{r_{ps}}{Y_{ps/s}}$$

The relationship between the two forms of the rate of substrate consumption is shown in *Figure 13*.

3.3 Product formation

The kinetics of product formation are imperfectly understood. Basically there are two kinds of products apart from the cell itself. These are indicated in Figure 12 and are:

(i) Products which are the result of energy production by the cell i.e., CO_2, water and simple carbon compounds such as ethanol, lactic acid, etc.

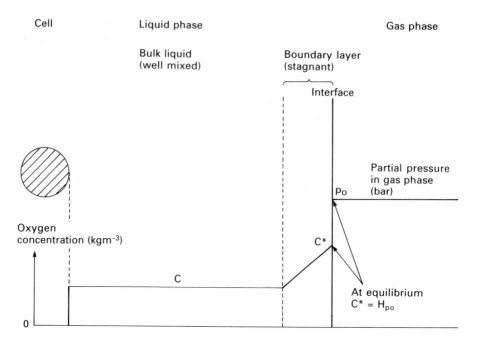

Figure 14. Physical model for mass transfer across a phase boundary.

(ii)   Products which require energy for synthesis such as polysaccharides, antibiotics, exo-enzymes etc.

For the first type of product Luedeking and Piret showed experimentally (and it is possible to show theoretically) that

$$r_{ps} = \alpha r_x + \beta x_v$$

where α and β are parameters.

The second type is more difficult to model. The production of complex products which require energy for synthesis depends on so many factors, e.g. cell age, inducers, precursors, etc., that there is no widely accepted theory. For our purposes it is sufficient to accept the simple first-order kinetic expression.

$$r_{pc} = \beta x_v$$

This type of product formation kinetics is called non-growth associated. If product formation rate is directly proportional to cell growth rate, i.e.

$$r_{pc} = \alpha r_x$$

then this is called growth associated. The Luedeking–Piret model is a mixture of growth associated and non-growth associated terms. This is called a mixed model.

The distinction between production of complex products and simple breakdown products is not normally made in modelling since it is usually clear from the context what we are dealing with. If the associated yield factor has a prime, then we are dealing with a complex product requiring energy for synthesis. If it is a yield coefficient without the prime then we are dealing with an energy producing breakdown product.

3.4 Transfer across phase boundaries

The general model for transfer of a substance across a gas/liquid or solid/liquid phase boundary is similar to Ohm's Law for electron transfer across a resistance, in that the flow is assumed to be directly proportional to the potential driving force across the resistance.

For mass transfer in fermentation systems, the resistance to mass transfer is assumed to be in a thin liquid film at the phase boundary called the boundary layer and the driving force is assumed to be the concentration difference of the material that is being transported across this liquid film. Obviously the total flow will depend on the area of the interface. In this case we write

$$N_{io} = k_L a \, (C^* - C)$$

where

N_{io}	is the flow	kg m^{-3}h^{-1}
$C^* - C$	is the driving force usually in concentration units	kg m^{-3}
a	is the interfacial area	m^2 m^{-3} = m^{-1}

and

k_L	is the mass transfer coefficient	m h^{-1}

Notice that the flow of material across the interface is based on unit volume of the medium, as is the interfacial area.

The driving force $C^* - C$ is the difference in concentration across the boundary layer. The assumed physical system is shown in *Figure 14*. C^* is the equilibrium concentration of the substance being transferred at the interface. This value has to be determined by experimental means or from the equilibrium saturation curve of the solute in the medium. Note that the concentration gradient is assumed to be linear in the boundary layer and constant in the bulk liquid. The presence of metabolizing microorganisms is assumed not to affect these assumptions. The combination of k_L and a is referred to as the volumetric mass transfer coefficient and is regarded as a single parameter, $k_L a$, with dimension h^{-1}. The value of $k_L a$ obviously depends on the value of a, the thickness of the boundary layer and the resistance to the diffusion of the solute through the boundary layer. These depend in a complex way on the hydrodynamics of the fermenter and the physical properties of the medium and its constituents. Standard textbooks should be consulted for values of $k_L a$.

4. MODELS

In the previous sections we have shown how to write the mathematical models of a simple batch fermentation and of a general system, and how to use the model to get values of some of the fermentation parameters. In this section we deal with several specialized models of the fermentation process. These are all derived from the generalized balance equation given in Section 2.1, which is reproduced below for the general quantity y,

$$\frac{d(Vy)}{dt} = (F_i \, y_i + V \, r_{gy} + V \, N_{iy}) - (F_o \, y + V \, r_{cy} + V \, N_{oy})$$

and

$$\frac{dV}{dt} = F_i - F_o$$

together with rate expressions for the rates of generation and consumption of y (r_{gy} and r_{cy}) and for the rates of transfer of y across the phase boundaries into and out of the liquid phase (N_{iy} and N_{oy}).

The specialized models are derived by dropping or modifying various terms in the balance equation and by our selection of the quantities, y, which we consider important. For example, if we make the accumulation terms zero, i.e., $d(Vy)/dt = 0$ *and* $dV/dt = 0$, we are dealing with a steady state continuous flow fermenter (chemostat). If we make F_i and F_o zero then we have a batch fermenter; with F_i finite and F_o zero we have a conventional fed-batch fermenter. If y_i depends in some way on y then we are dealing with a recycle fermenter and if we had several fermenters in series with the output of one being the input to the next then we would add an extra subscript to the variable indicating which fermenter in the series we were dealing with and write a balance equation for each one. Thus y_{1i} and y_1 would refer to the first, y_{2i} and y_2 would refer to the second and so on for all the variables F_o, F_o, V, r_{gy} etc., in all the fermenters.

We have already indicated that we have to decide which quantities are important in our fermentation system. For example in the yeast fermentation, as dealt with in the introduction, we decided that the viable cell concentration, x_v, the carbon substrate, S, and the nitrogen substrate, N, were important, and so we had three balance equations. If we were dealing with a fragile cell such as an animal cell line which was harvested for extraction of product, we might be interested in viable cells, x_v, dead cells x_d, and oxygen level, C, and consider all soluble nutrients to be in excess. Commercial aerobic fermentations are usually oxygen limited so that x_v and C would definitely be included.

Obviously the number of combinations of fermenter type (batch, continuous, etc.) and fermentation type (yeast, animal cell, oxygen limited, carbon dioxide sensitive, etc.) is very large and only a small selection can be covered. In the remainder of this section we deal with a few of the more common models. Each part will be laid out identically, with the balance equations and principal assumptions and applications first. This will be followed by some solutions of the equations in graphical form, a short discussion of the results and, in some cases, methods for determining parameters. To save repetition, the rate equations and parameter values are produced once only in the next subsection and will be referred to only in terms of the rate term in the individual models. Similarly the nomenclature and units are given only once at the beginning of the chapter.

4.1 Rate equations and parameter values

4.1.1 *Rate equations*

The rate equations used in the rest of this section are listed below:

Cell growth	r_x	$= \mu x_v$			
	where μ	$= \dfrac{\mu_m S}{k_s + S}$			(1)
	or μ	$= \dfrac{\mu_m C}{k_o + C}$			(2)
	or μ	$= \dfrac{\mu_m N}{k_n + N}$			(3)
Cell death	r_d	$= k_d x_v$			(4)
Product formation	r_p	$= \alpha r_x + \beta x_v$	(mixed)		(5)
		$= \alpha r_x$	(growth associated)		(6)
		$= \beta x_v$	(non-growth associated)		(7)

Carbon substrate uptake

aerobic $\qquad r_s = \dfrac{r_x}{Y'_{x/s}} + \dfrac{r_p}{Y'_{p/s}} + m_s x_v \qquad\qquad (8)$

anaerobic $\quad r_s = \dfrac{r_x}{Y'_{x/s}} + \dfrac{r_p}{Y'_{p/s}} \qquad\qquad\qquad (9)$

Oxygen substrate uptake $\qquad r_o = \dfrac{r_x}{Y'_{x/o}} + \dfrac{r_p}{Y'_{p/o}} + m_o x_v \qquad (10)$

Nitrogen substrate uptake $\qquad r_n = \dfrac{r_x}{Y_{x/n}} \qquad\qquad\qquad\qquad (11)$

Oxygen transfer gas to liquid

$$N_{io} = k_L a (C^* - C) \qquad\qquad\qquad (12)$$

4.1.2 Parameter values

The following parameter values are used in all the models discussed below except where values are given in the text. The units are kg, m, h and are given in detail in *Table 1*.

α	4.4	β	(EtOH) 0.035	β	(polysaccharide) 0.1		
C^*	0.007	γ	10.0	k_d	0.06	$k_L a$	100
k_n	0.1	k_o	0.00007	k_s	1.67	m_o	0.064
m_s	0.06	V_o	20	V_m	40	$Y'_{p/s}$	0.75
$Y_{ps/s}$	0.51	$Y_{x/n}$	6.84	$Y'_{x/o}$	1.5	$Y'_{x/s}$	0.8
$Y'_{x/s}$	0.51	μ_m	0.294				

4.2 Batch fermenter models

A typical aerobic yeast fermentation without oxygen limitation was described in the introduction to this chapter, Section 1.2. There is therefore no need to repeat this model. In this section we deal with two constant volume batch fermentations. The first is one in which the product is the breakdown product of the carbon source where it is anaerobically metabolized to provide energy for the cell processes. This product may be thought of as ethanol produced by *Saccharomyces cerevisiae* in a typical alcohol fermentation. The second is a typical non-growth associated excreted product such as polysaccharide produced under conditions where nitrogen is limiting.

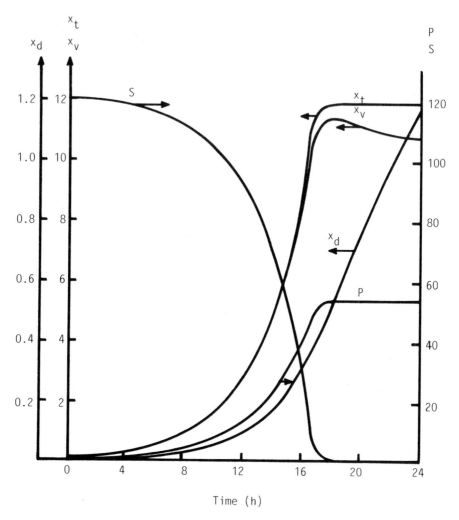

Figure 15. Model prediction for constant volume batch with anaerobic product.

4.2.1 *Constant volume anaerobic batch fermenter with anaerobic product*

(i) *Balance equations*

Viable cells
$$\frac{dx_v}{dt} = r_x - r_d$$

Non-viable cells
$$\frac{dx_d}{dt} = r_d$$

Anaerobic product
$$\frac{dP}{dt} = r_p$$

Carbon substrate
$$\frac{dS}{dt} = -r_s$$

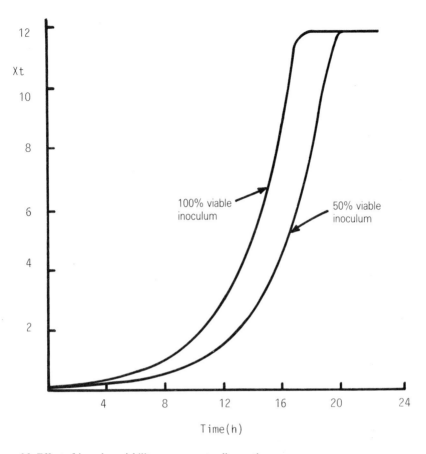

Figure 16. Effect of inoculum viability on apparent cell growth curve.

(ii) *Rate equations*

r_x (1), r_d (4), r_p (5), r_s (9) (numbers refer to the equations in Section 4.1)

(iii) *Total cell dry weight as measured experimentally*

$$x_t = x_v + x_d$$

The plots of the five variables versus time are shown in *Figure 15*. In this we assumed a 100% viable inoculum, i.e. an inoculum with no dead cells which had been harvested from a vigorously growing culture and had not been allowed to stand. Note how the non-viable cells gradually increase as a proportion of the total and how after the growth stops, because of exhaustion of substrate, the total dry weight remains constant although the viable cell concentration continuously declines.

If the inoculum were of only 50% viability we would get the behaviour illustrated in *Figure 16* where we have compared the apparent growth curves (x_t, total cell dry weight) for the 100% viable inoculum and a 50% viable inoculum. As you can see the apparent effect of the low viability is to give what appears to be a lag phase of approximately 2 h and a fermentation time which is similarly longer. This is not a true lag, although sometimes mistaken for one. A true lag period occurs when the organism has to adapt its internal enzyme spectrum to a production medium quite different to the inoculum medium. This can be modelled but the model is rather complex.

Models are only an abstraction from reality and this one is no exception. An important criticism is that it does not incorporate any mechanism for providing maintenance energy after the carbon source has run out. In reality the cells would consume their own substance (and shrink in size) by a mechanism known as endogenous respiration. Alternatively dead cells could lyse releasing nutrients into the medium sufficient for maintenance of the remaining viable cells. It is also not unusual for death rates to increase after the limiting substrate is consumed. All of these phenomena can be modelled but the models become rather complicated.

(iii) *Estimation of parameters*

The manipulations for determining those parameters not already covered in Section 1.1 are listed below.

Since $x_t = x_v + x_d$, $\dfrac{dx_t}{dt} = r_x$

Since $\dfrac{dP}{dt} = r_p = \alpha r_x + \beta x_v$

then $\quad \dfrac{1}{x_v} \dfrac{dP}{dt} = \alpha \dfrac{1}{x_v} \dfrac{dx_t}{dt} + \beta$

which give α and β. After growth stops we have $r_x = O$ so that

$$\frac{dx_v}{dt} = -r_d = -k_d x_v$$

which has the analytical solution

$$\ln (x_v) = A - k_d t$$

where A is an arbitrary constant. Thus a plot of $\ln(x_v)$ versus time after the carbon substrate has run out should give a straight line of slope $-k_d$. Note viability data will be required to determine x_v.

4.2.2 *Constant-volume aerobic batch fermenter, polysaccharide product*

In the previous model the rate of product (ethanol) formation was given by:

$$r_p = 4.4 \, r_x + 0.035 \, x_v.$$

Although a mixed model, it behaves almost like a growth-associated one because of the relative values of the coefficients. In this example we have a non-growth associated product formation model in which $r_p = 0.1 \, x_v \, S/(\gamma + S)$ which is typical of some extra-cellular polysaccharide producers. The factor $S/(\gamma + S)$ is included to ensure that r_p approaches zero as S approaches zero. In this type of fermentation the nitrogen source is the limiting substrate. This ensures that the cell concentration is restricted to such a value that oxygen transfer does not become the limiting factor in product excretion ($k_L a$'s are quite small in these viscous solutions). It also ensures that carbon substrate used for cell synthesis and maintenance is a reasonably low proportion of the total carbon source consumption. Cell death is ignored in the main model but its effects are illustrated later.

The model is as follows:

(i) *Balance equations*

Viable cells $\qquad\qquad\qquad\qquad\qquad\dfrac{dx_v}{dt} = r_x$

Limiting nitrogen substrate $\qquad\qquad\dfrac{dN}{dt} = -r_n$

Polysaccharide product $\qquad\qquad\dfrac{dP}{dt} = r_p \dfrac{S}{\gamma + S}$

Carbon substrate $\qquad\qquad\qquad\dfrac{dS}{dt} = -r_s$

(ii) *Rate equations*

$$r_x \ (3), \ r_n \ (11), \ r_p \ (7), \ r_s \ (8)$$

Plots of the four variables versus time are shown in *Figure 17*. The interesting point to note in this fermentation is the behaviour after the limtiing substrate has disappeared. At this point the viable cell concentration remains constant and the product increases linearly with time until the carbon substrate is beginning to be exhausted. This linear increase is typical of non-growth associated product formation. The depletion of the carbon substrate is related to the increase in polysaccharide and the constant $(m_s \, x_v)$ requirement for cell maintenance. As carbon substrate approaches zero, the rate of product formation and substrate consumption decrease due to the influence of the term $S/(\gamma + S)$ in the product formation model.

(iii) *Estimation of parameters.*
In the constant x_v region, we have:

$$\frac{dP}{dt} = r_p = \beta \, x_v . \frac{S}{\gamma + S}$$

Estimation of β and γ is by means of a Michaelis−Menten type plot as for μ_m and k_s. *Plot* $x_v/(dP/dt)$ versus $1/S$ to give a straight line of slope γ/β and intercept $1/\beta$.

The effect of cell death on the production of the non-growth associated product is easily modelled by including a term $(-k_d \, x_v)$ on the right-hand side of the viable cell balance. The effect of x_v and P is shown in *Figure 18* for this particular system at various values of k_d. The effect on the batch time to substrate exhaustion is quite dramatic. Times to reach this point which is also the point of maximum polysaccharide concentration are as follows:

$$k_d \ = \ 0 \quad 0.01 \quad 0.02 \quad 0.03 \ h^{-1}$$
$$Time \ = \ 40 \quad\ \ 46 \quad\ \ 58 \quad >80 \ h$$

4.3 Continuous fermenter models

A steady state well-mixed continuous fermenter is one into which the medium flows continuously and from which a mixture of spent medium, cells and products also flows continuously. The rate of inflow and outflow are steady. In the volume balance

$$\frac{dV}{dt} = F_i - F_o$$

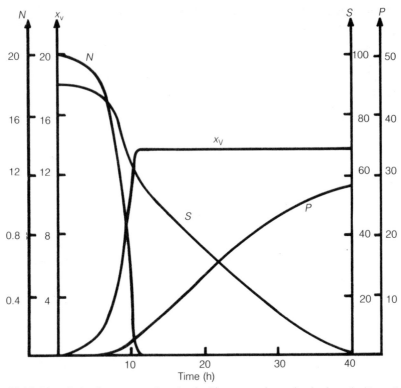

Figure 17. Model prediction for constant volume batch with non-growth associated polysaccharide production.

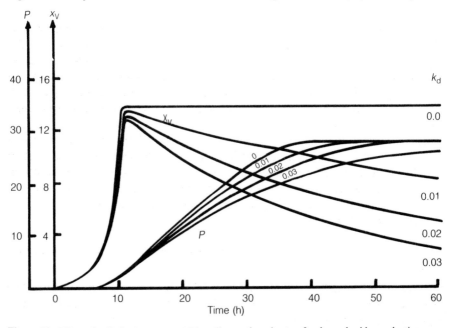

Figure 18. Effect of cell death rate on viable cell growth and rate of polysaccharide production.

F_i and F_o are equal so that the derivative of the volume, dV/dt, is zero; this is another way of saying that V does not vary with time. An additional requirement is that all values of the independent variables, usually the inlet concentrations (media composition) and the inlet flow, and all the environmental variables such as temperature and pH have been constant for such a period of time that all other variables (usually the exit concentrations and rates) are steady. This is called the steady state.

The general balance equation as outlined in Section 2.1 is

$$\frac{dVy}{dt} = (F_i \, y_i + Vr_{gy} + V \, N_{iy}) - (F_{oy} + V \, r_{cy} + V \, N_{oy})$$

Since $F_i = F_o = F$ and V is constant, it can be simplified to

$$\frac{dy}{dt} = \frac{F}{V} \, (y_i - y) + r_{gy} - r_{cy} + N_{iy} - N_{oy}$$

If we ignore transfer across the phase boundary and consider two dependent variables, viable cell concentration, x_v, and limiting carbon substate, S, then we have two equations which can be written:

$$\frac{dx_v}{dt} = D(x_{vi} - x_v) + r_x$$

$$\frac{dS}{dt} = D(S_i - S) - r_s$$

where $D = F/V$. These, together with the rate equations for r_x (1), and r_s (8) (note $r_p = 0$ in (8)), constitute a model which can be solved for any starting values of x_v and S given fixed values of D, x_{vi} and S_i and, of course, fixed parameters μ_m, k_s, $Y'_{x/s}$ and m_s.

Figure 19 shows some of the solutions of these equations for various starting points. No matter where you start from, you always end up at the same values of x_v and S after sufficient time has elapsed. These are called the steady state values for the defined values of the independent variables x_{vi}, S_i, and D. The dilution rate used in solving the above equations was 0.25 h^{-1}. The mean residence time of fluid flowing through the fermenter is $(1/D) = 4$ h. It takes between three and four mean residence times for the system to reach steady state.

At steady state neither x_v nor S is changing so that dx_v/dt and dS/dt must both be zero. Thus, in order to find the solution of the equations, all the derivatives in the balance equations are set equal to zero and the resulting set of algebraic equations is solved. If it is difficult to solve these algebraic equations, and sometimes it is so, it is possible to get round the problem by solving the differential equations and letting them run to steady state as in *Figure 19*.

4.3.1 *Steady state continuous fermenter for cell production with a single limiting substrate in solution*

This is the simplest possible model to start with. We assume no death of cells, a limiting substrate dissolved in the feed, all other nutrients including oxygen in excess and no cells in the inflowing medium.

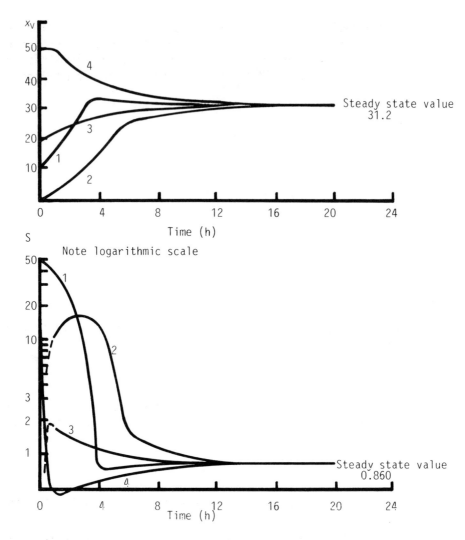

Figure 19. Approach to steady state for a continuous fermenter.

(i) *Balance equations*

Viable cells $\qquad O = r_x - D x_v$

Carbon substrate $\quad O = -r_s + D(S_i - S)$
$\qquad\qquad$ where $D = F/V$

(ii) *Rate equations* $\quad r_x$ (1), r_s (8), note $r_p = O$

Substituting the rate equations into the balance equations and rearranging gives:

$$D x_v = r_x = \frac{\mu_m S x_v}{k_s + S}$$

$$D(S_i - S) = \frac{r_x}{Y'_{x/s}} + m_s x_v$$

These two equations can easily be solved for x_v and S in terms of D, S_i, and the parameters. Note that this ease of solution only occurs when it is possible to cancel out one of the variables as in the first equation which is derived from the viable cell balance. Since there is a solution $x_v \neq 0$, we can cancel x_v in the first equation to give:

$$D = \frac{\mu_m S}{k_s + S}$$

so that

$$S = \frac{k_s D}{\mu_m - D}$$

In the second equation by substituting $r_x = D\,x_v$, we get

$$D(S_i - S) = D\,\frac{x_v}{Y'_{x/s}} + m_s\,x_v$$

so that

$$x_v = \frac{D(S_i - S)}{\dfrac{D}{Y'_{x/s}} + m_s}$$

$$= \frac{Y'_{x/s}D}{D + m_s\,Y'_{x/s}}\,(S_i - S)$$

Since we have S from the first equation we therefore have an expression for x_v in terms of the independent variables.

It is also important to list the constraints which are:

$$x_v \geq 0, \quad 0 \leq S \leq S_i$$

This simple analytical solution of the equation relies on the cancellation of x_v in the first equation. If this is not possible, e.g. if there were cells in the input stream so that the viable cell balance were

$$0 = r_x + D(x_{vi} - x_v)$$

then the solution of the equations would become complicated.

The solutions of these equations are plotted in *Figure 20* as full lines. The important points to note are:

(i) The outlet limiting substrate concentration is independent of the input limiting substrate concentration.

(ii) At a fixed dilution rate, called the critical dilution rate, D_c, the cell concentration drops to the constraint $x_v = 0$ and the limiting substrate concentration reaches the upper constraint $S = S_i$. Beyond this critical dilution rate cells are said to be washed out, since they are leaving at a higher rate than they are growing. The critical dilution rate is given by the expression $D_c = \mu_m S_i/(k_s + S_i)$. Since S_i is usually very much greater than k_s, D_c is approximately equal to μ_m.

(iii) The cells always grow at a specific rate μ equal to the dilution rate D. This can be seen from the viable cell balance which is $r_x = \mu\,x_v = D\,x_v$. Thus a chemostat, as it is often called, can be used to fix the growth rate of cells.

(iv) The general shape of the curves is fixed by the values of the parameters k_s and m_s. k_s determines the curvature of the falling part of the x_v and the rising part of the S curve as they approach the critical dilution rate. The smaller k_s, the

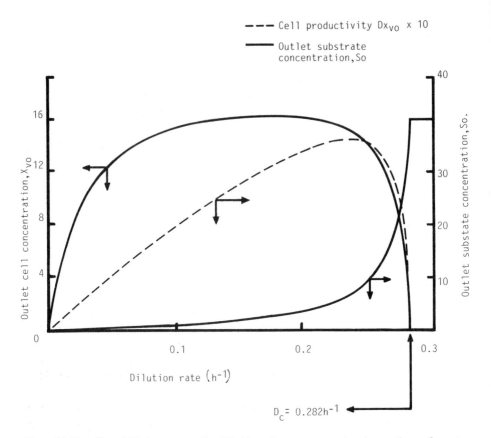

Figure 20. The effect of dilution rate on cell and limiting substrate concentrations in a continuous fermenter.

sharper the curvature. m_s determines the curvature of the rising part of the x_v curve at low dilution rates. The smaller m_s, the quicker the rise. When m_s is zero the curve starts from a maximum value of x_v equal to $Y'_{x/s} S_i$.

(v)   The maximum value of x_v is roughly equal to $Y'_{x/s} S_i$.

(iii) *Productivity.* The productivity of a batch culture is given by the final concentration of whatever is being made divided by the complete time of the batch (the running time and the emptying, cleaning, sterilizing and refilling time). The units are kg product $m^{-3} h^{-1}$. For the continuous fermenter there is no emptying, cleaning, sterilizing, and refilling component. Indeed this is one of its advantages. The productivity of such a fermenter is given by multiplying the dilution rate, D. by the concentration of the product in the outlet stream. In this particular case the cell productivity is Dx_v kg cells $m^{-3} h^{-1}$, and is plotted as the broken line in *Figure 20*.

4.3.2 *Steady state continuous fermenter with product formation*

We can easily extend the model of the previous example to take account of product formation by adding a balance equation for the product which if the input stream contains no product is

$$0 = r_p - DP$$

where P is the outlet product concentration. We can use the standard rate equation for product formation, equation (5). The substrate uptake equation, (8), for aerobic fermentation will be used. As before we have assumed a single limiting carbon substrate in solution and excess of all other nutrients.

The solutions of the algebraic equations describing this system are as follows:

from the cell balance:
$$S = \frac{k_s D}{\mu_m - D}$$

from the substrate balance:
$$x_v = \frac{D(S_i - S)}{\dfrac{D}{Y'_{x/s}} + \alpha D + \dfrac{\beta + m_s}{Y'_{p/s}}}$$

from the product balance:
$$P = \frac{(\alpha + \beta) x_v}{D}$$

These equations have been solved for two sets of conditions. *Figure 21* is for non-growth associated product formation ($\alpha = 0$) and *Figure 22* for mixed product formation kinetics using similar parameters to those used for *Figure 15*. The parameters were chosen to give roughly the same maximum productivity in each case. *Figure 20* serves as a growth-associated example since in that model product concentration would be directly proportional to cell concentration.

The general shape of the cell concentration and carbon substrate concentration versus dilution rate curves are the same in all three cases as illustrated in *Figure 20*. The principal effects of the differing kinetics are seen in the product concentration curve. For growth associated it follows the cell concentration curve; for non-growth associated it starts from a high value (equal to $\beta Y'_{p/s} S_i/(1 + m_s Y'_{p/s}/\beta)$) and drops sharply to zero as dilution rate increases to the critical value. For mixed kinetics the curve lies somewhere between the two. In the case illustrated (*Figure 22*) it is almost flat until close to the critical dilution rate. In all cases the productivity curve (*DP* versus *D*) has the same shape, rising steadily from zero and peaking at a dilution rate close to the critical.

In deciding whether to use continuous fermenters for product formation, the kinetics are very important. In particular you must consider both the concentration and the productivity of the product. Generally non-growth associated products are required in high concentrations to reduce recovery costs. Unfortunately the productivity is low at the highest concentration, and fed-batch fermenters are more suitable for this type of product. With mixed and growth-associated products it is possible to run continuously since high concentrations and high productivities occur at the same dilution rate.

4.3.3 *Oxygen limitation in continuous culture*

Oxygen as a limiting substrate is different in its implications to limiting substrates dissolved in the feed. Because of its very low solubility in water ($\simeq 0.007$ kg m^{-3} at

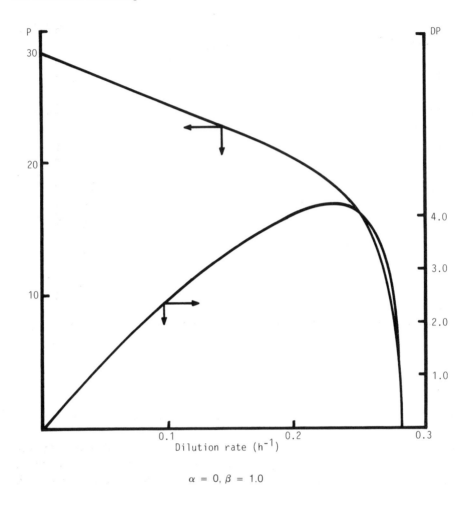

$$\alpha = 0, \beta = 1.0$$

Figure 21. Non-growth associated continuous product formation.

around 30°C) it is not possible to supply sufficient oxygen in solution to oxidize the carbon source and it must be transferred from the gas phase. The rate of transfer is limited by the volumetric transfer coefficient and the partial pressure of oxygen in the gas phase, p_o, and the rate of supply to the medium is independent of the flow rate of medium into the fermenter.

As a simple example to illustrate the problem we take the case already covered in Section 4.3.1 and add a balance equation for oxygen to take account of the fact that oxygen is now the limiting substrate. The model is therefore:

Balance equations

Viable cells	$O = r_s - D x_v$
Carbon substrate	$O = -r_s + D(S_i - S)$
Oxygen substrate	$O = -r_o + D(C_i - C) + N_{io}$

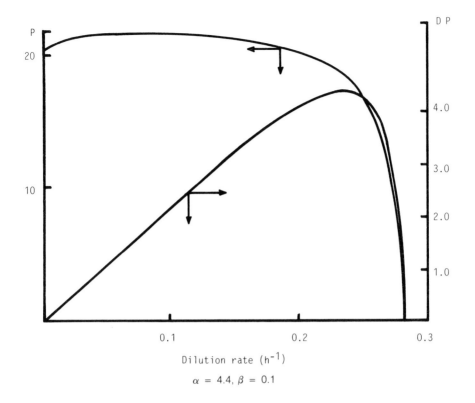

$$\alpha = 4.4, \beta = 0.1$$

Figure 22. Mixed growth and non-growth associated product formation in a continuous fermenter.

where $D = F/V$ and N_{io} is the rate of oxygen transfer into the medium from the gas phase. The rate equations when oxygen is the limiting substrate are: r_x (2), r_s (9), r_o (10), note $r_p = O$ in both these, and N_{io} (12). The usual constraints apply together with $O \le C \le C^*$.

The solution of these equations proceeds in the usual manner. From the cell balance, taking the non-zero value for x_v (so that it can be cancelled out) we get:

$$C = \frac{k_o D}{\mu_m - D}$$

From the oxygen balance, remembering that $r_x = Dx_v$, we get:

$$x_v = \frac{Y'_{x/o}(D(C_i - C) + k_La(C^* - C))}{D + m_o Y'_{x/o}}$$

and finally from the carbon substrate balance we get:

$$S = S_i - \frac{x_v}{DY'_{x/s}} (D + m_s Y'_{x/s})$$

The expression for x_v can be simplified if we assume, as is reasonable, that $D(C_i - C) << k_La(C^* - C)$ (k_La is of the order of $100\times D$) to give:

$$x_v = \frac{Y'_{x/o} k_La(C^* - C)}{D + m_o Y'_{x/o}}$$

105

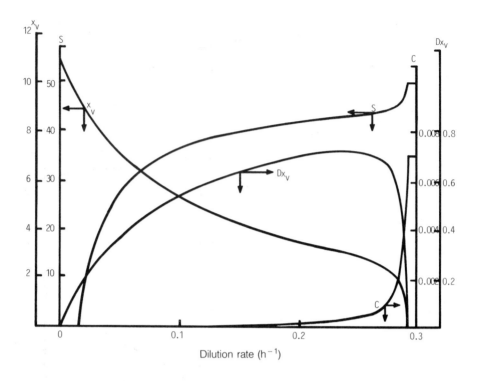

Figure 23. Oxygen limitation in continuous culture.

These three equations are plotted in *Figure 23*. The *C* versus *D* curve is exactly what we would expect for a limiting substate, but the plot of x_v versus *D* is quite different. This is because instead of having the variable *D* in the numerator of the expression we have the constant quantity $k_L a$. Thus we get a hyperbolic drop in x_v. The rate at which x_v drops depends upon the value of $k_L a$. The smaller the value of $k_L a$ the faster the value of x_v drops as dilution rate increased. Carbon substrate is in excess and depending on the value of S_i we get greater or less amounts flowing out in the outlet stream. The cell productivity is also severely limited by the oxygen transfer rate.

At low dilution rates the cell concentration is relatively high. This means that up to a dilution rate of 0.017, all the carbon substrate has been consumed which in turn means that our original assumption that oxygen is limiting is incorrect in this region, and in fact these high cell concentrations could not be achieved. The easiest way to analyse this situation is to determine another set of expressions for x_v, *S*, and *C*, on the assumption that carbon substrate is limiting. At any given dilution rate we then choose to use that limitation, oxygen or carbon substrate, which gives us the lowest value of predicted cell concentration, x_v. The rate equation for carbon substrate limitation is

$$r_x = \frac{\mu_m S}{k_s + S} x_v$$

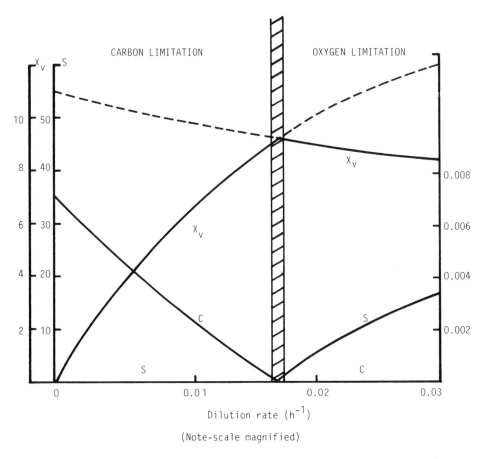

Figure 24. Transition between carbon substrate and oxygen substrate limitation in continuous culture.

which when substituted together with the other rate equations into the three balance equations gives, after some manipulation:

$$S = \frac{k_s D}{\mu_m - D}$$

$$x_v = \frac{D Y'_{x/s} (S_i - S)}{D + m_s Y'_{x/s}}$$

and $C = C^* - \dfrac{(D + m_o Y'_{x/o}) x_v}{k_L a \ Y'_{x/o}}$

The results for x_v, C, and S in the restricted range of dilution rate between 0 and 0.03 h^{-1} are plotted in *Figure 24* to show what happens. Below a dilution rate of 0.017 h^{-1}, the fermentation is carbon limited and S is practically zero. As D increases the cell concentration rises on the normal curve but the oxygen concentration falls from its saturated value of 0.007 kg m^{-3} until at a dilution rate of between 0.017 and 0.018 h^{-1} it has become limiting, and in turn almost zero. At this point the cell concentration now begins to fall on the curve for oxygen limitation, shown dashed in the carbon limited region. Cell growth is now restricted by oxygen and the carbon substrate concentration rises, since the inflowing carbon is in excess of that required for growth and energy production.

This switch has important implications, particularly in biological waste treatment where oxygen limitation is the rule and where the object of the fermenter is to have a low carbon substrate concentration in the effluent. The point at which the switch occurs is given approximately by the expression:

$$k_L a \ C^* \ = \ (Y'_{x/s}/Y'_{x/o})D \ S_i$$

which allows a quick estimation of any one of $k_L a$, C, S_i, or D given values of all the others.

4.4 Fed-batch fermenter models

Neither constant-volume batch fermenters nor continuous fermenters are suitable for non-growth associated products. The problem is that in order to achieve high concentration of the products we need to have high concentrations of cells in the fermenter for long periods of time and, as we have shown previously, this is not possible with constant volume batch or continuous fermenters. One possible solution would be to maintain a high concentration of cells in a continuous fermenter by immobilizing cells on an inert support and passing the production medium over them. This technique is used in some cases but there are problems in maintaining the viability of the cells on the support. The analysis of these systems is beyond the scope of this chapter. The other solution which is common in the fermentation industries is first to build up a high concentration of cells in the batch fermenter and then to switch the metabolism of the cell so that it produces the desired biochemical. This switch may be effected by stopping cell growth and adding precursors of the biochemical and making other enviromental changes as necessary. If growth were merely stopped by stopping the supply of the carbon source, then the cells would immediately start to die at a high rate. What is done is to feed carbon and oxygen at a rate sufficient for the maintenance energy needs of the cells and for synthesis (including the energy necessary for synthesis) of the biochemical. If too much carbon is fed, the cells will grow and this may switch their metabolism but in any event it would mean that the oxygen transfer capabilities of the fermenter would be exceeded and, amongst various other problems, the fermentation would go anaerobic.

This type of fermentation is called fed-batch. There are two phases, a growth phase in which the cells are grown to the required concentration and then a production phase in which carbon source and other requirements for production are fed to the fermenter in a relatively concentrated liquid stream at a precalculated rate. Eventually because of death of the cells the production slows down (or alternatively the fermenter would overflow) and at this point the batch is harvested. If the microorganism is particularly robust, the solera system may be used. In this method a part of the batch is harvested at intervals and feeding is continued, as in sherry production.

The model considered is very simple. We assume in the first instance that cells do not die, that oxygen is always in excess, and that the limiting substrate is the carbon source and that any other nutrient requirements either in the growth or production phase will be supplied in sufficient quantity. The model is in two parts:

(i) *Growth phase, constant volume.* The usual batch equations apply.

Cell balance
$$\frac{dx_v}{dt} = r_x$$

Carbon substrate balance
$$\frac{dS}{dt} = -r_s$$

Rate equations r_x (1), r_s (8), note $r_p = 0$

The initial conditions are $x_v(0)$ and $S(0)$ and the usual constraints apply. At the end of the growth phase S will be zero and x_v will be x_{vb}.

(ii) *Production phase.* The fermentation is no longer constant volume since carbon substrate is being fed into the fermenter. The model equations are now

Cell balance
$$\frac{dVx_v}{dt} = Vr_x = 0 \qquad \text{(no growth)}$$

Carbon substrate balance
$$\frac{dVS}{dt} = -V\, r_s + FS_i = 0 \qquad \text{(since } S \text{ must be zero for } r_x = 0\text{)}$$

Product balance
$$\frac{dVP}{dt} = Vr_p$$

Volume balance
$$\frac{dV}{dt} = F$$

Rate equations
$$r_s = m_s\, x_v + r_p/Y'_{p/s}$$
$$r_p = \beta\, x_v$$

Note that carbon substrate is fed into the fermenter at a rate F m^3 h^{-1} and a concentration S_i, that substrate is used only for cell maintenance and product synthesis, that the carbon substrate concentration in the fermenter must be kept equal to zero by some external control system, and that although the total amount of cells in the fermenter does not change the concentration does since the volume is increasing.

The initial conditions for this production phase are $x_v = x_{vb}$, $S = 0$ and $V = V_0$. An additional constraint is that V must be less than V_m, the maximum working volume of the fermenter.

The growth phase equations are solved by normal numerical methods for the solution of differential equations. The production phase has an analytical solution which is as follows for constant S_i:

$$x_v = x_{vb}/(1 + k\, x_{vb}\, t)$$

$$\text{where } k = \frac{1}{S_i}\left(m_s + \frac{\beta}{Y'_{p/s}}\right) \text{ and } t = \text{time}$$

$$P = \beta\, x_v$$
$$V = k\, V_0\, x_{vb}\, t$$
$$F = k\, V\, x_v$$

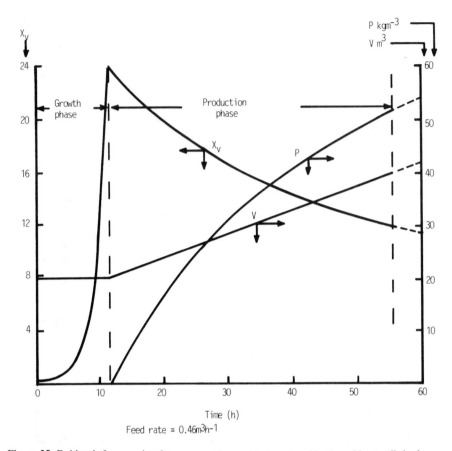

Figure 25. Fed-batch fermentation for non-growth associated product kinetics with no cell death.

These results are plotted in *Figure 25* for V_o equal to 20 m^3 and S_i equal to 200 kg m^{-3}. The batch was assumed to be harvested when the total volume had doubled to 40 m^3. The initial growth phase lasted for 11.5 h. At this point the carbon source in solution was fed at a rate of 0.46 m^3 h^{-1}. The cell concentration gradually declines due to the dilution effect of the feed. The feed rate remains constant because the total amount of cells in the fermenter does not change and therefore the carbon requirement for maintenance and product formation does not change. The volume increases linearly as does the total amount of product in the fermenter although the product concentration does not follow the linear path because of the dilution effect. The fed-batch phase lasts for 43.5 h, making a total of 55 h for the fermentation and the final product concentration is 51.7 kg m^{-3}. This gives a total product of 2068 kg and the fermenter achieves an overall productivity of 37.6 kg h^{-1} (excluding non-fermenting time).

Death of cells or any conversion to a non-productive form reduces productivity. To model this effect we change the cell balance to:

$$\frac{\mathrm{d}Vx_v}{\mathrm{d}t} = -k_d\, x_v$$

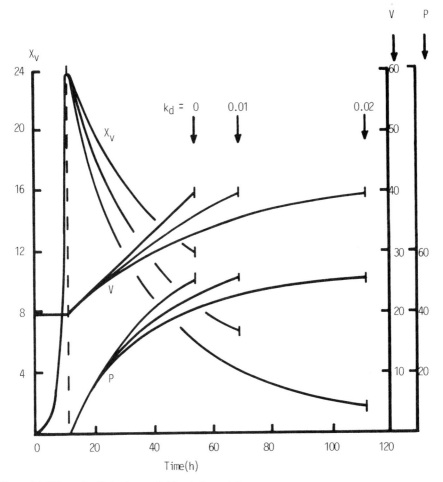

Figure 26. Effect of cell death on a fed-batch fermentation.

where k_d is the death rate constant. The effect on the production phase of varying k_d is shown in *Figure 26*. It has been assumed for simplicity that the viability is 100% at the beginning of this phase.

Since the total amount of viable productive cells is declining, the feed rate must decline to compensate and this shows up in the volume versus time curve which is no longer a straight line. The decline in productive viable cells extends the length of the fed-batch phase but has no effect on the final product concentration or amount produced if the full 40 m^3 is used. The effect on total fermentation time and productivity can be drastic, as shown below:

Death rate k_d, h^{-1}	0	0.001	0.005	0.01	0.02
Total fermentation time, h	55	56.5	60.5	68.5	102.5
Productivity of batch, kg h^{-1}	37.6	36.6	34.2	30.2	20.2

111

As fed-batch fermentation is of such economic importance, models of the type illustrated are of great importance for planning production strategy. By incorporating an oxygen balance and providing the kinetics are known, the true behaviour of the fermentation can be modelled with reasonable accuracy. It is then possible to simulate various strategies and to optimize the length of the growth phase, the maximum cell concentration, the feeding patterns, the harvest time, etc.

5. AFTERWORD

In this chapter we have considered the mathematical modelling of some simple fermentation systems. It is important to emphasize the limitations of the chapter.

We have dealt only with well-mixed systems in which the microorganisms are freely suspended. We have considered all substrates other than oxygen to be readily soluble in the liquid medium. We have only considered the one case of a substrate being transferred across a phase boundary (oxygen) and we have not considered the transfer of other gases out of the liquid medium into the gas phase.

The treatment of the kinetic rate equations has been very elementary, and we have made no mention of the effect of environmental conditions upon the values of the kinetic parameters. Our treatment of mass transfer processes has been very basic.

Although freely suspended microorganisms are still very common in industry there are many developments in immobilizing microorganisms either as films on solid surfaces or as freely suspended aggregates in inert supports such as open structure foam particles or entrapped in a solid matrix. The treatment of these systems, although it starts from the same premises as we have outlined in this chapter, leads to different mathematical forms which are more difficult to solve. In particular we end up with partial differential equations instead of the ordinary differential equations of this chapter, and instead of simple models for transfer across a phase boundary we have to take account of diffusion in a continuum.

Finally we should emphasize that the development of the subject by mathematicians and engineers depends crucially upon the insights into the biological mechanisms which are discovered and developed by the microbiologists, biochemists, and molecular biologists to whom we owe the very foundations of the subject.

CHAPTER 5

Fed-batch and continuous culture

ANNE BROWN

1. INTRODUCTION

A variety of techniques has been developed for the submerged culture of micro-organisms. Batch and fed-batch culture have been part of the art of fermentation since antiquity and are still used today for most industrial fermentations. Continuous culture techniques are relatively new and have not yet found much industrial application but have proved invaluable as a tool for laboratory studies.

There are few descriptions in the literature of the methodology of submerged culture due, perhaps, to the diverse range of fermentations and the vast amount of equipment on the market. This chapter will describe some of the general aspects of setting up and running fed-batch and continuous culture, and will compare these with simple batch culture.

2. FED-BATCH CULTURE

A fed-batch culture is essentially a batch culture which is supplied with either fresh nutrients, for example growth-limiting substrates, or additives, such as precursors to products. The terms 'extended culture' and 'semi-batch' have also been used, on occasion, to describe this culture technique.

The fed-batch culture technique was originally devised by yeast producers in the early 1900s to regulate the growth in batch culture of *Saccharomyces cerevisiae* (bakers' yeast), with malt as a substrate. The yeast producers faced two problems: first, if the malt concentration in the culture medium was high then the yeast, instead of producing biomass, produced ethanol. Second, if the concentration of malt was kept to a minimum then yeast growth was restricted. The problems were solved by a controlled feeding regime (usually in hourly increments) so that yeast growth remained substrate-limited.

Following the successful application of fed-batch techniques to yeast production modifications of the technique were then applied, in the early 1940s, to the production of glycerol, butanol, acetone, and organic acids with great success, i.e. higher product yields and more efficient utilization of medium constituents (see (1) for an extensive review of these processes).

In more recent years fed-batch culture techniques have been used for the production of antibiotics (for example penicillin G) vitamins, amino acids, enzymes and growth hormones (1,2).

The basic theory of fed-batch culture is covered in Chapter 4.

Table 1. The classification of fed-batch culture techniques

Technique		Example			Reference
Status	Method	Control parameter	Substrate/ additive	Product	
With feedback control	Indirect	RQ	Molasses	Yeast	(3)
	Direct	Ethanol	Ethanol	SCP	(4)
Without feedback control	Intermittent or incremental additions	None	Phenylacetic acid	Penicillin	(5)
	Constant rate additions	None	β-Galactosidase	Glycerol	(6)
	Exponential rate additions	None	Ethanol	SCP	(7)

2.1 Control techniques for fed-batch culture

Numerous fed-batch culture control techniques have been devised. For simplicity, the techniques have been classified into two main groups: those with feedback control and those without. The two groups are further subdivided (*Table 1*).

Indirect feedback control involves monitoring fermentation parameters which are indirectly related to the substrate which may or may not be limiting, e.g. dissolved oxygen, respiratory quotients (i.e. CO_2/O_2 ratios) and pH values. Direct feedback control involves directly monitoring the concentration of substrate in the culture broth. In addition, the substrate can be either kept at a constant value or can be varied to maintain an optimum concentration. Both feedback control techniques can either be carried out manually (open-loop system) or automatically (closed-loop system). In systems without feedback control the feed can either be added at a constant rate (e.g. a fixed value on a pump setting), added exponentially in accordance with the biomass increase, or simply in increments (e.g. hourly). (Chapter 7).

2.2 Advantages of fed-batch culture

It offers a means of:

(i) utilizing substrates which inhibit growth if present at high concentrations (e.g. ethanol, methanol, acetic acid, and aromatic compounds);

(ii) increasing the production of biomass (e.g. in a batch culture of yeast, the biomass may reach $5-10g$ litre^{-1} whereas in a fed-batch culture the biomass may increase ten-fold). This aspect is greatly favoured in the production of growth-associated products (Chapter 4);

(iii) producing secondary metabolites which are not growth-related (i.e. those produced when the organism enters its stationary phase). In this case the feed rate can be controlled to initially produce high biomass then, as the organism enters stationary phase and growth slows down to finally give just enough substrate for maintenance energy whilst product formation is taking place (see Chapter 4);

(iv) overcoming catabolite repression;

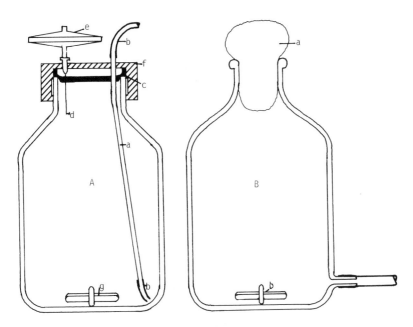

Figure 1. Holding vessels. A. Screw-neck borosilicate glass vessel with medium/inoculum addition assembly. (a) Stainless steel rod; (b) silicon tubing; (c) silicon disc; (d) hypodermic needle; (e) air vent; (f) screw cap; (g) magnetic bar. B. Aspirator-type vessel for introducing an inoculum of filamentous fungi into the fermenter. (a) Cotton-wool plug; (b) magnetic stirrer bar.

(v)   reducing broth viscosity (e.g. in the production of dextran and xanthan gum;

(vi)   overcoming the problem of contamination, mutation and plasmid instability which are found in continuous culture.

2.3 Disadvantges of fed-batch culture

(i)   Additional instruments for feedback control may be costly.

(ii)   In systems without feedback control, where the feed is added on a pre-determined fixed schedule, it is difficult to deal with any deviations in the organism's growth pattern (i.e. time courses may not always follow expected profiles).

(iii) It requires a substantial amount of operator skill.

2.4 Ancillary equipment

2.4.1 *Holding vessels*

All the holding vessels required for a batch culture will be required for a fed-batch culture. The holding vessels, in particular those for acid and base, must be constructed from a non-toxic, corrosion-resistant material which is capable of withstanding repeated sterilization cycles. Borosilicate glass (Pyrex or Duran) falls into this category, and has the added advantages of being both inexpensive and transparent. Stainless steel vessels are also suitable, but costly. A suitable grade of stainless is 316 (AISI) (equivalent to 585 (EN) and 4336 (Workstoff)).

Figure 1 illustrates two methods of assembling holding vessels for easy transfer of

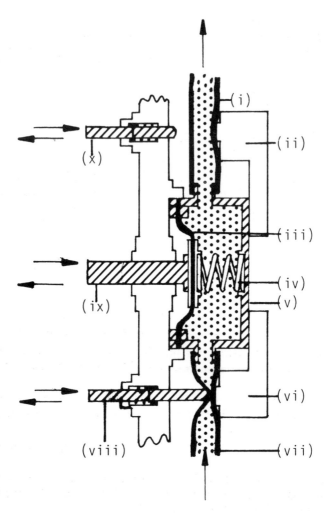

Figure 2. Operating principle of a diaphragm dosing pump. (i) Pressure discharge tube; (ii) upper pressure plate; (iii) diaphragm; (iv) return spring; (v) pump head; (vi) lower pressure plate; (vii) suction intake tube; (viii) suction valve plunger; (ix) diaphragm plunger; (x) pressure valve plunger.

liquid additives to the culture vessel. Both types A and B can also be used to introduce inoculum into the culture vessel. Type B is more suitable for filamentous fungi because it has a wide bore outlet.

As a guideline the following vessel sizes are sufficient for 10 litre fermentation. 2 × 1 litre (acid and base); 500 ml (antifoam); 1 litre (extra nutrient additions); 500 ml (inoculum). In addition a vessel for holding the feed will be required—a 10 litre Duran should be sufficient.

2.4.2 *Pumps*

There are two types of pump (both positive displacement, i.e. for every stroke or revolution they displace a quantity of fluid) which are suitable for the aseptic pumping of small volumes of culture media: the peristaltic pump and the diaphragm-dosing pump.

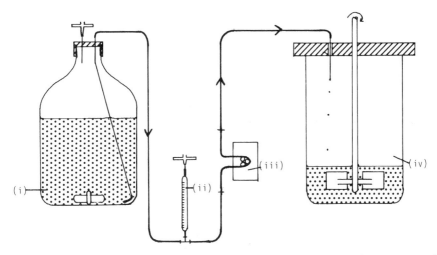

Figure 3. Schematic diagram of a fed-batch culture system. (i) Feed reservoir; (ii) flow meter assembly; (iii) peristaltic pump; (iv) STR.

Other pumps, such as vane, piston, and centrifugal types are unsuitable because they are difficult to sterilize and cannot be used for pumping small volumes.

(i) *The peristaltic pump.* The peristaltic pump (discussed in Chapter 1) consists of a main body for housing both the drive motor and electrics and rotating unit of rollers (usually three or six).

In operation the advancing roller occludes the tube which, as it recovers to its original size, draws in fluid that is trapped by the next roller. The fluid is then expelled as the unit moves round.

The flow rate can be varied by either the speed setting or by changing the bore diameter of the tube.

A suitable pump for most fed-batch operations is the Watson–Marlow 503S/R (100 r.p.m.) which gives flow rates from $0.2-4.2$ ml/min^{-1} with a 0.5 mm bore tube and $0.6-12.0$ ml/min^{-1} with a 0.8 mm bore tube.

(ii) *The diaphragm-dosing pump.* A diaphragm-dosing pump suitable for both fed-batch and continuous culture is the Braun FE211. The pump consists of a main body and a detachable heat-sterilizable head. The operation of the pump is illustrated in *Figure 2*. Fluid is sucked in via the suction inlet tube (vii) to the pump head (v) (the pressure discharge tube (i) being closed). The suction inlet tube then closes and the pressure discharge tube opens forcing fluid out. The suction and pressure forces in the pump head are generated by the reciprocating action of both the diaphragm plunger and the return spring. The diaphragm (made of Viton) is extremely durable and can survive at least $8-10$ heat sterilization cycles before needing to be replaced.

The advantages of the pump are that it is capable of accurate metering at low flow rates; it does not require frequent calibration and it can act as its own check valve against back-flow of fluid. The main disadvantage of the pump is that (unlike the peristaltic pump) it has limited application (i.e. to obtain a wide range of flow rates different pump heads have to be purchased, and they are costly), although one suitably chosen head should give a sufficient flow rate range to carry out most fed-batch processes.

2.5 **The fed-batch process**

The lay out of a typical fed-batch proces is illustrated in *Figure 3*.

A fed-batch culture is first initiated as a simple batch culture. The inoculum for the batch culture can be prepared by the shake-flask technique, which is one of the simplest methods of culturing an organism. The technique is essentially a small batch culture with typical lag, exponential, and stationary growth phases. The lag phase in the main batch culture must be kept to a minimum, and therefore the physiology of the cells in the inoculum is of prime importance. In most cases the inoculum should be cultured in a medium which is similar to that in the main culture vessel. The physical conditions should also be similar (e.g. in temperature and pH). The inoculum should not be transferred into the main vessel until the cells are growing in the exponential phase (this takes approximately 12 h). The inoculum can then be transferred under aseptic conditions to a holding vessel, then pumped using a peristaltic pump (Watson-Marlow or equivalent) into the main culture vessel. A 10% (v/v) quantity of inoculum should suffice. For a fed-batch culture the batch culture volume is usually half that of the final volume.

Careful attention should be paid to the start-up of the fed-batch operation, especially with substrate-limited processes which use feedback control (see Section 2.1). In such cases the feed should be started immediately after all the substrate in the batch culture has been utilized, otherwise the process may be difficult to control. Control problems may also arise if the cutlure is allowed to 'starve' before starting the feed, resulting in a long lag phase whilst the cells alter their metabolic state.

A rough guide as to when to start feed is when a sharp rise in dissolved oxygen is noted.

2.5.1 *Determination of feed flow rate*

The following can be used to calculate the feed flow rate for a substrate limited fed-batch process:

$$FS_{\mathrm{o}} = \frac{\mu X}{Y}$$

where *F is the feed flow rate (litre* h^{-1}*)*, S_{o} is the concentration of limiting substrate in the feed (g litre^{-1}), μ is the specific growth rate (h^{-1}), X is the biomass concentration (g litre^{-1}) and Y is the yield of cells on substrate (g biomass per g substrate).

2.5.2 *Pump calibration*

In order to maintain accurate control of feed rate, frequent flow rate checks are necessary. This is particularly true for peristaltic pumps, since tubing in the pump head tends to become worn over long periods of operation, resulting in inaccurate flow rates.

For determining flow rates an in-line flow meter can be installed. A simple flow meter can be easily constructed out of a 10 mL graduated pipette (see *Figure 3*). Under normal working conditions the tubing to the flow meter is closed with a gate clip. To calibrate the pump, simply open the clip and allow the pipette to fill (it should be situated below the liquid level in the feed reservoir) then close off a clip to isolate the feed reservoir. After isolating the feed reservoir the flow rate can be determined by measuring the level drop in the pipette over a given time interval.

3. CONTINUOUS CULTURE

Continuous culture techniques have been used on an industrial scale for over 50 years. Some of the earliest uses include vinegar production, yeast propagation, and waste treatment, such as the trickling filter and activated sludge processes.

Laboratory-scale continuous culture studies are comparatively new. They are a useful tool for studying the growth and physiology of microorganisms. Extensive research has been carried out on both laboratory- and industrial-scale continuous culture systems, and the literature on the topic is vast. The best overall view can be obtained from eight successive publications of the Continuous Culture Symposia; Volume 8 (8) provides a clear perspective on what practical and research uses are being made of the technique.

Continuous culture is basically a method of prolonging the exponential growth phase of an organism in batch culture. The technique involves feeding the organisms with fresh nutrients and at the same time removing spent medium plus cells from the system in such a fashion that several factors remain constant with time, i.e. culture volume, cell concentration, product concentration, and culture environment (e.g. pH, temperature, dissolved oxygen). This condition is known as steady state. Providing the culture remains free from contamination and the organism is stable (i.e. not subject to spontaneous mutations) then the system can be operated for long periods of time.

The main advantage of such a system is that cell mass and other products can be produced under optimal environment conditions. On an industrial scale this technique is more desirable than batch or fed-batch because it is significantly more productive in terms of fermenter down-time per unit of product manufactured.

3.1 Control techniques for continuous culture

There are several control techniques which can be applied to continuous culture. Substrate limitation, i.e. the chemostat culture technique, is the most extensively studied and for this reason the relevant practical aspects have been emphasized in this chapter. A second technique which is studied to a lesser extent because of its limited application is the turbidostat. Some other techniques which are less well studied but worthy of mention are the nutristat, pH-auxostat, CER-stat, DO_2-stat and OUR-stat.

3.1.1 *The chemostat*

In the chemostat the feed medium is designed to contain, in excess, all but one of the nutrients essential for growth. The particular nutrient which is not in excess effectively determines the rate at which the organism grows. Several equations have been derived to describe the behaviour of the chemostat in steady state; readers are advised to refer to Chapter 4, which explains the equations in greater detail.

When a chemostat is in steady state there is a mass balance across the system, i.e. the flow of feed medium into the vessel equals the flow of culture out. The formation of new biomass in the culture medium is a result of the addition of fresh nutrients and is balanced by the loss of biomass in the outflow, consequently the biomass in the vessel remains constant.

The flow of medium into the vessel (F) is related to the culture medium (V) by the term dilution rate (D), which is defined as $D = \dfrac{F}{V}$. The flow rate is expressed

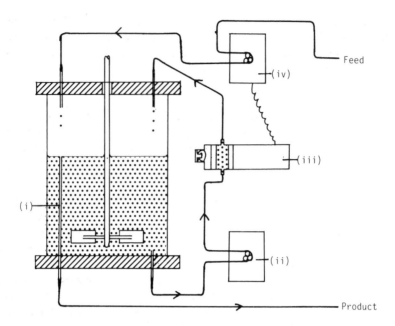

Figure 4. Schematic diagram of the operation of a turbidostat. (i) Overflow weir; (ii) peristaltic pump (circulatory); (iii) photometer/control module; (iv) peristaltic pump (feed).

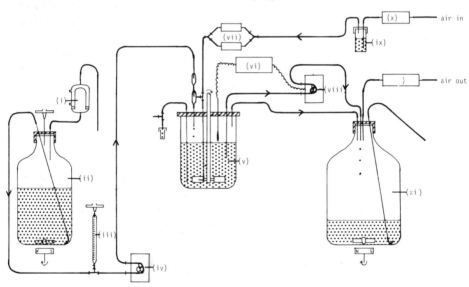

Figure 5. Schematic diagram of a laboratory chemostat culture. (i) Medium filter; (ii) feed reservoir; (iii) flowmeter assembly; (iv) peristaltic pump (feed); (v) STR; (vi) level control module; (vii) double air filters; (viii) peristaltic pump (product); (ix) air humidifier; (x) air filter; (xi) product receiver.

in litres h^{-1}. The volume is expressed in litres and as a result the dilution rate is expressed in units per hour.

Under steady state conditions the biomass in the vessel remains constant, therefore the specific growth rate (μ) must equal the dilution rate, i.e. $\mu = D$. The equation below has been derived from the material balance equation for the limiting nutrient across the system. From the equation the relationship between yield of cells on substrate ($Y_{x/s}$) and limiting substrate can be calculated:

$$X = Y_{x/s} (S_o - S)$$

where X is the biomass at steady state (g litre^{-1}), S_o and S are the concentrations of limiting substrate in the feed and residual substrate in the outflow, respectively (g litre^{-1}). $Y_{x/s}$ is the yield of cells on substrate (g biomass per g of substrate).

3.1.2 *The turbidostat*

In this system, the feed contains an excess of all the required nutrients so growth is not substrate-limited and, as a result, the organism grows at its maximum specific growth rate (μ_{max}). The system is controlled at a desired cell density by constantly monitoring optical density. This can be achieved using a continuous-flow photoelectric cell in a recirculation loop as shown in *Figure 4*.

Optical density can either be measured by a nephelometer (scattered light detector) or spectrophotometer (transmitted light detector). During operation optical density is continuously monitored. If the monitor detects a deviation from a set-point value then a signal is relayed to the controller which adjusts the rate at which the feed is added to the vessel.

The turbidostat is particularly useful in cases where nutrient limitation is not desirable such as the degradation of toxic wastes and the selection of antibiotic resistant mutants. For further reading on turbidostats and their uses refer to reference (9).

Unfortunately the technique has two major inherent disadvantages, namely cell growth on the sides of the photocell and gas bubbles entrained in the circulating medium. These result in inaccurate optical density measurements, and consequently the system is frequently difficult to control.

Laboratory-scale turbidostats are commercially available from New Brunswick Scientific.

3.2 Ancillary equipment

A chemostat culture can be run with a few modifications to basic batch culture equipment. The following section gives instructions on how to carry out such modifications and lists, where applicable, factors which influence choice of equipment. Reference should be made to *Figure 5* which shows the overall layout of a typical laboratory single-stage chemostat.

3.2.1 *Holding vessels for feed medium and product*

To run a small volume chemostat (e.g. 2 litre) at low dilution rates, 10 litre and 20 litre capacity Duran bottles (or equivalent) should be sufficient to hold the feed medium and the product. *Table 2* should be used as a guideline as to how much feed is required per day for different culture volumes at various dilution rates.

Assemble the vessels as described in section 2.4.1. A silicon bung, with sufficient inlet holes, strapped to the neck of the bottle with cable ties, will be more secure for long-term operation than the disc assembly.

The feed will have to be replenished daily. Providing all the nutrients are soluble they can be filter-sterilized directly into the feed reservoir via a filter capsule assembly. Suitable filter assemblies are commercially available (Pall, Millipore, Sartorius).

Incorporate a dip-tube (6 mm) into the harvest receiver assembly so that medium can be drawn off as the vessel fills. A stainless steel connecting rod, suitably placed, should be incorporated in line so that the receiver can be replaced if need be.

3.2.2 *Pumps*

Two pumps are usually required to operate the chemostat, but this depends on the type of level control mechanism installed (see section 3.2.5). Suitable feed pumps have already

Table 2. Guide to the daily feed requirements of a laboratory chemostat

Working culture volume	Litres of feed required at dilution rate			
	0.05	0.1	0.25	1.0
2	2.4	4.8	12	48
5	6.0	12	30	120
10	12	24	60	240

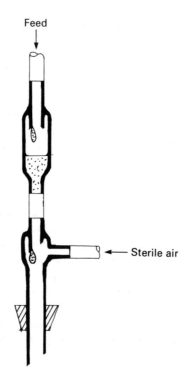

Figure 6. Drip-feed assembly.

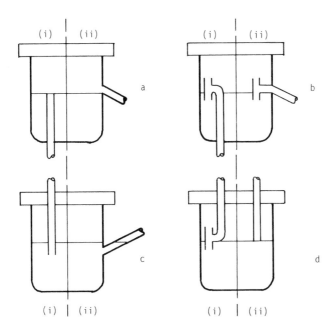

Figure 7. Methods of level control. (**a**), (**b**), Overflow weirs; (**c**), (**d**), pumping-out mechanisms.

been discussed in section 2.4.2. To operate the overflow a peristaltic pump capable of producing a faster flow rate than the feed pump should be used. The Watson-Marlow 101UR or equivalent is suitable for this application. Some fermenters are supplied with two-channel, peristaltic pump modules (i.e. both inlet and outlet flow rates will be the same). The main disadvantage of such a system (depending on the level control mechanism) is that entrained gas in the outlet can lead to level control difficulties which are detrimental to achieving steady state.

3.2.3 *Drip-feed assembly*

During long-term chemostat operation the growth of cells back up the feed line is a major problem. The problem can be alleviated to a certain extent by incorporating a double drip-feed assembly into the feed line (*Figure 6*). The device serves to both prevent aerosols entering the feed-line and to act as a break in feed flow. A single drip-feed assembly can be constructed using two silicon bungs, two short lengths of stainless steel tubing, and a borosilicate glass tube (5.0 × 1.6 cm). A suitable double assembly can be purchased from Gallenkamp.

If growback still presents a problem, and providing the feed medium is heat stable, then an infra red lamp can be used to heat an appropriate section of the feed-line.

3.2.4 *Exhaust and aeration*

When operating with small culture volumes (i.e. 2−5 litre) at low dilution rates and relatively high temperature (i.e. 37°C), evaporation is a major problem especially if the fermentation is operated for a long period.

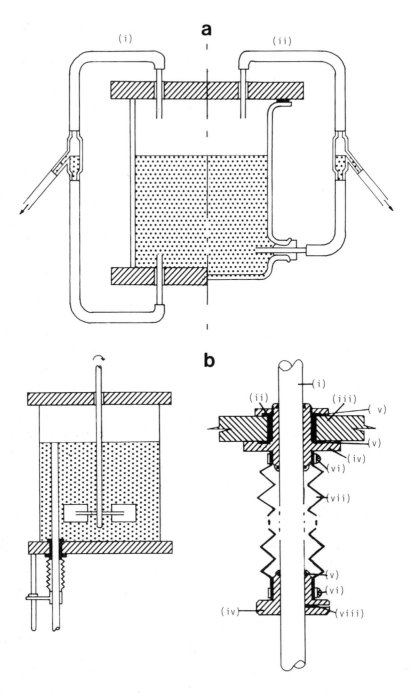

Figure 8. Adjustable weirs. (**a**) Adjustable external weirs: (i) fermenter with, and (ii) fermenter without base plate. (**b**) Adjustable internal weir: (i) stainless steel tube; (ii) collar retaining nut; (iii) vessel base plate; (iv) collar; (v) O-ring; (vi) jubilee clip; (vii) rubber bellows; (viii) locknut.

This problem can be overcome to some extent with an efficient condenser; however, it may be necessary to install an in-line air humidifier. A simple humidifier can be constructed out of a 1 litre Duran bottle (or similar vessel) as shown in *Figure 4*. It is necessary to install two in-line air filters so that, in case of failure due to blockage, the reserve filter can be brought into use without having to make any aseptic connections.

The exhaust gas should be channelled into the harvest receiver (this acts as a safety mechanism in case of a foam-out).

3.2.5 *Level control mechanisms*

In order to achieve and maintain steady state there must be some provision made for controlling the culture volume which is measured in litres of ungassed liquid. This presents a problem in aerobic cultures because the medium has to be aerated. A second problem, common to both aerobic and anaerobic cultures, is that as the culture environment has to be homogeneous, constant stirring is essential. Stirring in turn creates turbulence, which again leads to false level detection.

There are several methods of controlling culture volume; choice depends on (a) kind of organism (i.e. bacteria, yeast, or filamentous fungi), (b) type of fermentation (i.e. anaerobic or aerobic), (c) medium constituents (e.g. solubility and miscibility are important considerations), (d) culture volume, (e) degree of accuracy (5% volume variation is acceptable), (f) size of fermenter, (g) cost, (h) degree of operator skill.

Level control devices can be categorized into five groups, as follows:

(i) *Overflow weirs.* Overflow weirs (*Figures 7a* and *b*) are simple, inexpensive and easy to install. Problems can arise if foaming occurs resulting in a disproportionate number of cells being removed from the system; a T-piece can be installed at the outflow which will lower the take-off point if foaming is a problem. A further problem arises in fermentations which employ hydrocarbon substrates immiscible with water and less dense. In this case a disproportionate amount of substrate will be removed from the system.

Weir-type mechanisms are not suitable for filamentous fungi because the mycelium tends to aggregate around the top of the weir, eventually leading to blockage. The use of a (10 mm diameter) teflon weir in conjunction with a teflon-coated fermenter vessel has been put forward as a successful method of cultivating filamentous fungi (10).

Adjustable weirs external (11) and internal (12) (*Figures 8a* and *b*) allow for inbuilt flexibility should the liquid level rise during stirring. Such weirs are sterilized along with the fermenter, and no aseptic coupling is necessary. With the external weir tubing-length must be kept to a minimum, otherwise undesirable anaerobic pockets will develop.

(ii) *Pumping-out mechanisms.* With a pumping-out mechanism (*Figures 7c* and *d*) the culture volume is regulated to the depth at which the harvest tube is placed. If the take-off point is below the surface of the medium, then pumping out has to equal pumping in. This is difficult to achieve, but can be done with a two-channel peristaltic pump. The harvest tube must be introduced in an area of low turbulence where it is unlikely to draw in bubbles. Pumping-out mechanisms are easy to install and useful for small vessels where access through the base plate is not practical. They are more costly than weir-type systems.

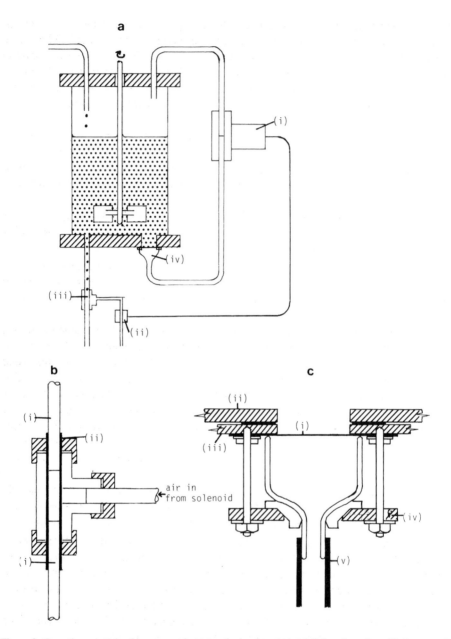

Figure 9. Operating principle of a manometric system for level control. (**a**) (i) Level monitor; (ii) air-actuated solenoid; (iii) pinch valve. (**b**) Pinch valve: (i) stainless steel tube; (ii) silicon tubing. (**c**) Pressure transducer: (i) flexible membrane; (ii) vessel base plate; (iii) backing plate; (iv) backing flange; (v) silicon tubing.

(iii) *Manometric systems.* The manometric system (*Figure 9*) was designed specifically for use with filamentous fungi. The system can be easily constructed and installed on any laboratory fermenter with an accessible base plate. Basically it is a U-tube manometer with one leg being the fermenter and the other being an external connection to the top plate. The increase in culture mass caused by feed flowing into the vessel will displace the flexible membrane of the pressure transducer (*Figure 9c*) and cause an increase in the fluid level in the external leg of the U-tube. This is detected by a level monitor (Fisons Limited) which triggers an air-operated pinch valve (*Figure 9b*), which in turn allows the culture to flow out until the original volume is attained (discharge volume is about 50 ml). (13).

The pinch valve (*Figure 9b*) consists of a piece of silicone tubing mounted inside a stainless steel T-piece and supported at either end by stainless steel rods.

The constriction is effected by compressed air and a normally open three-way solenoid valve (Schrader – Bellows Limited) prevents medium escaping until the manometer senses an increase in volume.

The main disadvantage is that the silicon tube and membrane are subject to wear and frequently have to be replaced. Marprene which has recently come on the market may be more reliable. The manometric system is available commercially.

(iv) *Level probe.* Conductivity and capacitance level probes are commercially available from most fermentation equipment suppliers. They are, however, costly and are not suitable for volumes of less than 5L. Furthermore, foaming can lead to false level detection.

A level control probe can either be used in conjunction with either a solenoid/pinch valve assembly (previously described) or a pumping-out mechanism. A level control probe which can be easily and cheaply constructed is shown in *Figure 10*. This probe consists of a stainless steel body (i) which houses an inner tube (ii) and a disc (iii). As long as liquid level is low, air flows down the outer tube and back up the inner tube. As the culture level rises the membrane (iv) deforms and presses the disc against the bottom of the inner tube causing a back pressure. The back pressure is detected by a transducer which signals the solenoid valve (vi) to close thus opening the pinch valve (v). Alternatively the transducer could send a signal to a pump (as shown in *Figure 5*).

This particular design of level probe is more accurate than capacitance or conductivity probes and therefore can be used with small culture volumes. The system can be used for filamentous fungi and tends to be more reliable than the manometric system.

(v) *Load cells.* Load cells are often fitted to large (>30L) base-mounted fermenters. A load cell is essentially an elastic body (e.g. a steel cylinder) which distorts when compressed. The amount of distortion (strain) is measured by a number of electrical resistance strain gauges which are fixed around the surface of the body. The changes of resistance with strain are proportional to the load on top of the cylinder. When the load (measured in kg) exceeds a set point value, a signal is sent either to start a pump or to open a pinch valve and discharge a given volume of culture.

Load cells have several drawbacks, the main one being cost. Furthermore, they are not very accurate (i.e. at low dilution rates steady state is disturbed) and therefore unsuitable for small culture volumes. Another disadvantage is that all tubing has to be supported to avoid inadvertant changes in weight (large vessels need flexible couplings), consequently the system is clumsy to operate. The main advantage is that as weight is the control parameter, foaming has little effect on level control.

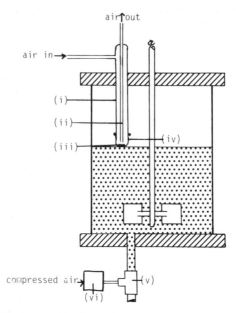

Figure 10. A simple level probe assembly: (i) stainless steel body; (ii) inner tube; (iii) disc; (iv) membrane; (v) pinch valve; (vi) solenoid valve.

3.3 The culture process

Set up the chemostat as for a batch culture. The batch culture medium should be half the strength of the feed medium: this reduces the oscillation effect seen in the transition phase between batch and continuous. During the transition phase the organism goes through a lag phase where it alters its nucleic acids, enzyme content and other structures to the new steady state level (for further reading on causes and prevention of oscillations during the transition phase refer to Dunn and Mor, reference 14).

At the start of the batch phase the culture volume should be at least 20% greater than the volume at which control takes place. The extra volume can then be drawn off to a predetermined level before the continuous phase is started.

Start the feed when the organism enters into the exponential growth phase. The dilution rate must not exceed the μ_{max} of the organism (μ_{max} if not already known from literature can be estimated from a simple batch culture experiment).

3.3.1 Determination of steady state

In practical terms steady state is achieved when biomass, substrate and product concentrations are constant with time. Theoretically, a chemostat will have reached steady state when the culture has undergone at least three residence or replacement times (six are more commonly used). The residence time (T_r) is the reciprocal of the dilution rate, i.e. $T_r = \dfrac{1}{D}$ (e.g. if $F = 0.5$ litres h^{-1} and $V = 2$ litres then $D = 0.25$h^{-1} and $T_r = 4$h).

Once steady state has been established, step changes in dilution rate must be small

Table 3. Results of growth rate relationships between contaminant/mutant populations (μ^c) and native populations (μ^n) in continuous culture

Growth rate relationship	Result
$\mu^c > \mu^n$	The mutant population rises exponentially and eventually replaces the native population
$\mu^c = \mu^n$	The mutant population rises at a constant rate
$\mu^c < \mu^n$	The mutant population assumes a constant value

otherwise oscillations will occur. At least three residence times should be allowed to elapse after making a step change in dilution rate.

3.3.2 *Problems associated with continuous culture*

(i) *Contamination and genetic instability.* Contamination and genetic instability (the latter a result of spontaneous mutation) in a culture may be a serious problem particularly in those cultures which are likely to run for long periods of time. For argument's sake contaminants and mutants can be treated as one because they have the same growth rate relationships with the original organism.

There are three possible outcomes with respect to the growth rate relationships between the original 'native' organism and mutant/contaminant organism (these are summarized in *Table 3*). In the first instance, if the specific growth rate of the contaminant/mutant is greater than of the 'native' organism then the contaminant/mutant will take over in an exponential fashion. This is a particular problem at low dilution rates. In the second instance, if the specific growth rate of the contaminant/mutant is equal to that of the 'native' organism then the new population will rise at a constant rate equal to its introduction. Contamination with such an organism may not be serious if the rate of entry or mutation rate can be kept negligibly small. In the third instance, if the growth rate of the contaminant/mutant is less than the 'native' organism then it assumes a constant value and only becomes a serious problem if its rate of entry into the culture is high. Sometimes media can be designed to reduce the growth rate of likely contaminants (e.g. low pH); it is, however, difficult to regulate the appearance of spontaneous mutants.

(ii) *Plasmid instability.* A plasmid is a self-replicating genetic element that confers traits non-essential for growth on the microorganism. The plasmid can be artificially constructed to express a foreign protein in its host cell (e.g. a bacterium such as *Escherichia coli* can be transformed to express insulin). The survival of a plasmid in a population depends on its ability to replicate within the host cell and subsequently pass into daughter cells. If this process does not occur then the number of cells with no plasmid in the population will rapidly outnumber the plasmid containing cells.

Plasmid instability can be controlled by giving the host cell a selective advantage. This can be done by incorporating, for example, an antibiotic-resistant gene into the plasmid so that only cells with the plasmid will survive in a culture medium containing, for example, an antibiotic. A second method is to delete a gene from the host chromosome which expresses a specific essential cell component. The host gene is then replaced on the plasmid and consequently cells with the plasmid will survive.

For further reading and methodology on plasmids, see reference (15).

4. ACKNOWLEDGEMENTS

The author wishes to thank Andrew Powell for his kind assistance with drawings and critical review of the manuscript.

5. REFERENCES

1. Yamane,T. and Shimizu,S. (1984) *Adv. in Biochem. Eng. Biotech., Vol. 30.* Fiechter,A. (ed.), Springer-Verlag, Berlin, p. 147.
2. Whittaker,A. (1980) *Process Biochemistry, 15,* 32.
3. Wang,H.Y., Cooney,C.L. and Wang,D.I.C. (1977) *Biotech. Bioeng., 19,* 69.
4. Huang,S.Y. and Chu.W.-B. (1981) *Biotech. Bioeng., 23,* 1491.
5. Singh,K. and Johnson,M. (1948). *J. Bact., 56,* 339.
6. Gray,P.P., Dunhill, P. and Lilly,M.D. (1973) *Biotech. Bioeng., 15,* 1179.
7. Yamane,T., Hirano,S. (1977) *J. Ferm. Technol., 55,* 380.
8. Dean,A.C., Ellwood,D.C. and Evans,C.G. (1984) *Continuous Culture.* **8,** Ellis Horwood, Chichester.
9. Munson,R.J. (1970) In *Methods in Microbiology. Vol. 2.* Norris,J.R. and Ribbons,D.W. (eds), Academic Press, London, p. 349.
10. Rowley,B.I. and Bull,T. (1973) *Lab. Prac., 22,* 286.
11. Herbert,D., Phipps,P.J. and Tempest,D.W. (1965) *Lab. Prac., 14,* 1150.
12. Callow,D.S. and Pirt,S.J. (1961) *J. Appl. Bact., 24,* 12.
13. Brown,D.E. and Inkson,M.B. (1972) *Biotechnol. Bioeng., 14,* 1045.
14. Dunn,I.J. and Mor,J.R. (1975) *Biotechnol. Bioeng., 17,* 1805.
15. Hardy,K.G. (1987) *Plasmids: A Practical Approach.* IRL Press, Oxford.

CHAPTER 6

pH, dissolved oxygen and related sensors

PETER J.HALLING

1. INTRODUCTION

The growth and behaviour of cells in a fermenter are obviously much dependent on the chemical composition of the environment that surrounds them in the culture medium. Knowledge of chemical concentrations is therefore of great value in understanding the course of the fermentation and their control may be necessary for reproducible operation. Clearly instruments that can continuously monitor chemical characteristics of the broth are desirable, but very few reliable sensors that can be directly inserted into the broth are available. (A major problem is the requirement to survive heat sterilization.)

The probes that are available, to measure pH and dissolved oxygen, are thus of great importance. The pH often tends to vary during culture growth and its control is usually desirable. Equally, too low levels of dissolved oxygen are a frequent cause of poor performance in aerobic fermentations. Hence both these probes would normally be considered essential for any use of fermenters other than routine growth of cells to a defined recipe (and even there they would provide a useful check!). In very low-oxygen or anaerobic cultures, the redox probe may be useful instead of a dissolved oxygen sensor. Most of the hints and tips in this chapter are given without citation. This does not imply that they are my own original discoveries. The majority have been passed on to me at various times by friends and colleagues who work with fermenters – the way most of us have learned the practice of laboratory fermentation. In many cases I cannot remember who told me, and in most I do not know who originated the idea. However, I would like to record my thanks to a series of friends and colleagues who have been generous with their time and knowledge, particularly in University College London, Unilever Research, PHLS Porton Laboratory and here at Strathclyde. I hope in turn that in the future this chapter can help some of those starting out in the fermentation game.

2. pH SENSORS

2.1 The equipment

2.1.1 Sensors

A number of manufacturers offer pH probes or 'electrodes' for fermentation (see Section 6). It is essential to use a probe designed for this purpose, so that it will survive the heat sterilization process. The sensors available are without exception combination pH probes, with both a glass electrode and a reference electrode, connected to the medium via a liquid junction at a small porous plug, usually on the side of the probe (*Figure 1*).

131

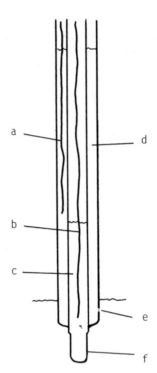

Figure 1. Typical design for a sterilizable pH probe. Showing the two internal compartments. (a) reference electrode; (b) inner electrode; (c) inner electrolyte; (d) reference electrolyte (refillable); (e) porous plug; (f) pH-sensitive glass.

The premier name is undoubtedly Ingold, who are therefore generally the most expensive source. I know of no careful study of the accuracy and/or reliability of Ingold probes compared with other cheaper makes—however, since the price difference is not enormous, many opt for Ingold on the basis of known historical performance. Usually different probes are offered depending on whether the fermenter is to be sterilized *in situ* or in an autoclave. If the former is used, then the electrode will need to be mounted in a special housing, usually supplied by the fermenter manufacturer, that allows its outside to be pressurized to more than 1 atm gauge during sterilization. This is to prevent the pressure inside the vessel forcing material back into and through the porous plug. For autoclave sterilization, special electrical connections may be provided to prevent problems due to their exposure to live steam in the autoclave.

2.1.2 *Meters and connectors*

The pH probe is a voltage-generating electrochemical cell. However, its internal resistance is very high (of the order of $10^9 \ \Omega$), so the potential generated can only be measured with the aid of a very high input impedance DC amplifier, which draws only a minute current. Suitable amplifiers are included in a wide range of available pH meters and controllers. In view of the low price of pH meters, it does not seem sensible for even the most impoverished of laboratories to try to construct its own. An instrument with a high current or voltage signal output is advisable, so that computer data logging is possible, either now or some time in the future. The high impedance of the probe also places stringent requirements on the connectors and electrical leads used between the sensor and the meter, so again it is wise to use a purpose-designed screened cable.

2.2 Use of the pH probe

2.2.1 *Calibration*

The probe must be calibrated immediately before use, as one of the last operations before sterilization of the fermenter. Calibration is performed with the sensor outside the fermenter, dipping it into suitable containers of one or more standard buffers. Ideally these should be at the temperature at which the fermentation will be run. The probe must obviously be connected to the pH meter that will be used during the fermentation and the calibration settings on this are adjusted by the same procedures as for any pH meter. Since recalibration later will be very difficult or impossible, it is wise to use some method to lock the calibration controls on the pH meter, to prevent accidental shifts during the experiment—many commercial instruments offer fixing screws for this purpose.

2.2.2 *Sterilization*

After calibration, the probes should be inserted and sealed into the fermenter. The connection to the meter will normally be removed during sterilization of the fermenter (certainly if an autoclave is used)—afterwards it can be reconnected and pH monitoring commenced. Some workers use ethanol sterilization of pH sensors separately from the fermenter, with the aim of extending probe life, particularly when using cheaper models. The probe must then be inserted and sealed into the vessel. I doubt whether this can be achieved aseptically, so it will probably lead to contamination unless the fermentation under study is particularly resistant to this. However, some do report regular use of the procedure without problems. For those who wish to try, the method involves placing the probe, plus a suitable fitting for mounting in the fermenter headplate, in absolute ethanol for at least an hour. The probe and fitting must be scrupulously clean and should be immersed to above the level of the latter. At the end of the process it should be transferred as rapidly as possible to the pre-sterilized fermenter, which should have the air supply already connected and flowing.

2.2.3 *Check of calibration*

It is possible that the calibration may shift as a result of sterilization or during use. Such changes should not be more than 0.2 units with a probe in good condition. However, some workers recommend calibration or recalibration after sterilization of the fermenter. Systems allowing fully aseptic partial withdrawal of the probe into calibration buffer have been designed for larger fermenters. On a laboratory scale, however, it is necessary to remove the probe completely and attempt to resterilize by chemical treatment as described above, which I would not normally recommend. A better means of checking a suspect calibration during an experiment is by removal of a broth sample (using the aseptic methods described in Chapter 8), measuring its pH outside the fermenter and comparing with the probe reading. If this approach is used, however, the external pH reading should be made as rapidly as possible after collecting the sample, as continued metabolism of the cells under now-changing conditions (e.g. oxygen depletion, nutrient depletion if in continuous culture) can give rapid changes in pH if the medium is poorly buffered, with significant alterations in minutes.

2.3 Routine maintenance

2.3.1 *Electrolyte*

The probes must be refilled or topped up with 'electrolyte solution' from time to time. This is in fact the electrolyte for the reference electrode and is slowly lost through the porous plug. Manufacturers will supply proprietary liquids for their probes, but some workers make their own: these are concentrated or saturated KCl solutions, sometimes with addition of further solutes to increase viscosity (usually recommended when the probe is to be pressurized for *in-situ* sterilization). In theory the level of electrolyte should be kept high enough such that its hydrostatic pressure exceeds that of the fermentation broth, so that any tendency to flow through the porous plug is from the probe into the culture. In practice this requirement is rarely met! Though the electrolyte level may be above that of the broth, the headspace of an aerated fermenter will normally be a little pressurized relative to the atmosphere, due to resistances in the exit gas filter and lines. The probe electrolyte compartment headspace, however, will normally be at atmospheric pressure during operation. As little as 0.01 bar of pressure difference would require 10 cm head in the probe, the likely maximum in practice. This may contribute to some of the problems with fouling of the porous plug, noted below. When the fermenter in use provides a sealed probe housing for pressurization during sterilization, it might be beneficial to keep this at a small positive gauge pressure during the fermentation, though I am not aware of any systematic test of this.

2.4 Problems in use

Problems likely to be experienced with pH probes are: (i) reduced sensitivity/slope; (ii) slow response; (iii) noisy signals; and (iv) mechanical damage.

2.4.1 *Slope/sensitivity*

This requires a little explanation. There is a theoretical relationship betwen pH and the potential between the probe electrodes, with the latter varying by $RT/F.\ln 10$ per pH unit (this slope is equal to 59 mV per pH unit at 25°C; R is the gas constant, T the absolute temperature and F the Faraday). A new pH probe will approximate to this theoretical slope, but reduced sensitivity is observed with ageing or damaged probes. Most pH meter/amplifiers will have a control that varies the sensitivity assumed in converting the voltage signal to pH — this is usually labelled slope or sensitivity and may be calibrated in mV or temperature (since the latter is the only variable that theoretically should affect slope). It is *not* the same as the 'set buffer' or 'zero' control(s). *Figure 2* illustrates what is observed if the sensitivity control is wrongly set. The system has been calibrated (with the set buffer control) at a single pH value as shown, and is then tested in one or more other buffers. Instead of the expected solid line, the meter reading will systematically deviate from the known buffer pH values in the manner

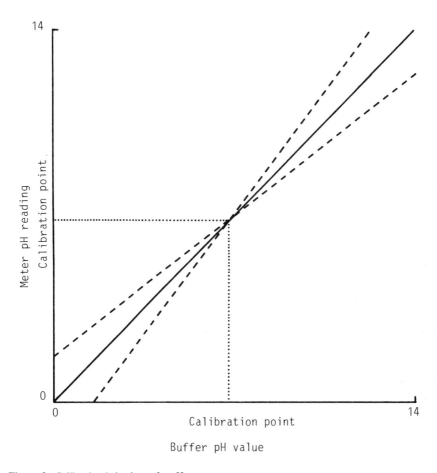

Figure 2. Calibration behaviour of a pH sensor.

shown by the dotted lines. If the steeper line is observed, the slope is set too low: if the shallower line, it is too high. The sign of poor sensitivity is that the slope/sensitivity control must be set to a value much less than theoretical to make the probe behave according to the solid line. (This is equivalent to making a two-point calibration of the probe, as described in the instructions for most pH meters or sensors.)

2.4.2 *Cleaning*

Should a pH probe show a lengthening response time and/or falling sensitivity, then cleaning may offer at least a partial reversal. A common cause of probe deterioration is contamination of the porous plug by material from the fermentation broths. This is often indicated by its staining brown or black – if it is clean, it should be white! Soaking in 10 mM HCl is a safe option, which does not appear to risk damage to the probe. (It is probably a good idea to routinely store the probe in this between runs.) Addition of pepsin is sometimes suggested to help remove protein deposits. If HCl treatment is not effective, two other possibilities may be tried, though both risk further damage to the probe. The probe may be soaked for an hour or two in roughly 1% H_2O_2 solution. Alternatively, gently mechanical cleaning of the porous plug may be tried — scraping with a sharp blade to remove deposits on the outer surface can be effective.

2.4.3 *Electrical interference*

The very high impedance of the pH probe and amplifier circuit can lead to problems. It makes the probe rather susceptible to noise from induced voltages due to stray fields from other electrical apparatus and to erroneous response due to minute current leakage between the two conductors carrying the probe signal. To limit these problems special screened leads and connectors will be supplied by the manufacturers of either the probe or the meter, but some difficulties may still remain. If excessive noise is observed, it may be possible to reduce it simply by moving the pH probe lead away from other electrical cables. The stirrer motor can be a source of interference, and this can be checked by turning it off for a few seconds. Spikes on the pH trace may coincide with a heater circuit switching on or off (this can usually be spotted by watching a light or listening for a relay changing in the heater control unit). Noisy or inaccurate readings after autoclave sterilization may reflect water contamination of the connectors and leads exposed to steam during this process. If so, some further protection by wrapping or taping may help.

2.4.4 *Avoiding mechanical damage*

The pH probe is physically rather fragile and prone to accidental breakage during setting up and cleaning of the fermenter. To minimize the risk here, it is advisable to insert the probe only at a late stage during preparation of the fermenter (it will need calibrating at this point anyway) and to remove it as the first operation in stripping down after use. (Many broken probes result from lifting the head plate with the sensor still in place!) To minimize the risk of damage during storage between runs, a convenient method is to place the probe in a plastic measuring cylinder containing the appropriate solution. Choose the size of the measuring cylinder so that the broader section of the probe rests on the rim, while the sensing bulb is suspended above the bottom. (A less desirable alternative is to put a cotton-wool plug at the bottom of the cylinder.) Where possible, clamp the measuring cylinder to some immovable structure.

3. DISSOLVED OXYGEN SENSORS

3.1 **Equipment**

3.1.1 *Design principles*

Dissolved oxygen probes for fermentation, often called 'oxygen electrodes', operate on one of two quite different principles. The majority are still of the 'galvanic' class, usually the 'Mackereth' or 'Borkowski—Johnson' designs. As the name suggests, these probes contain a galvanic cell (i.e. one generating an electric current), linking reduction of molecular oxygen at a noble metal (often silver) electrode with oxidation and dissolution of a lead electrode. The 'polarographic' (often 'Clark') type incorporates

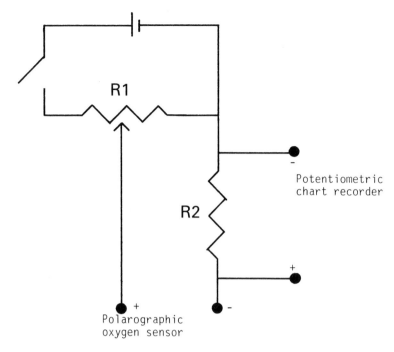

Figure 3. Circuit for use with a polarographic dissolved oxygen probe. R_1 and R_2 should be selected using the rules described in the text.

an electrolytic cell (i.e. one passing a current generated externally), in which oxygen is reduced at a platinum electrode. In each case the current flowing reflects the availability of oxygen at the surface of the noble metal electrode; usually it is determined simply by the rate at which oxygen arrives by diffusion. All the probes incorporate a membrane that separates the electrolyte and cell from the fermentation broth. This is permeable to oxygen and other gases, but not to the probe electrolyte and other ionic species. Oxygen diffuses across the membrane and the rate it does so determines the current passed by the probe circuit.

3.1.2 *Selection of probes*

It is essential to use a probe specifically designed for fermentation, and hence able to survive sterilization by moist heat. In particular this requires a membrane thick enough and properly supported in order to survive pressure differences across it that will be generated during the process (see the further comments in Section 3.2.1 below). Most of the available probes are offered for use with either autoclave or *in situ* sterilization and do not require any special treatment prior to sterilization (such as pressurization). In some cases a special moisture-resistant cable connection may be offered for autoclave sterilization, which is not necessary if sterilizing *in situ*. A number of different manufacturers offer oxygen probes (see Section 6), but in my experience most designs are not satisfactory. Regular problems of drift or complete failure to respond will be found and probe lifetimes are just a few months of regular usage. One model of probe,

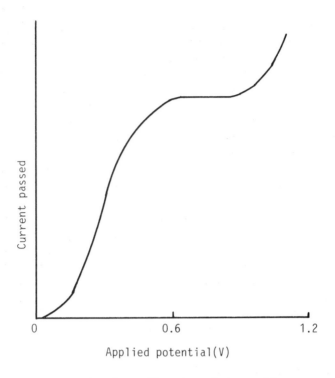

Figure 4. Electrical behaviour of a polarographic oxygen probe immersed in oxygen-saturated medium.

the polarographic type manufactured by Ingold, is greatly superior to all others I have tried. (These probes have the same design as a series offered by Instrumentation Laboratories Inc. in the USA.) This is reflected in the price, which is some 20 times that of the cheapest on offer. However, even at such a price, the probe cost is not a large proportion of the total for a typical laboratory fermenter. Hence unless the objective is just routine growth of cells, it seems pointless to economize by making dubious one of the most important measurements in any fermentation experiment. The Ingold probes seem to last at least several years in regular use (with occasional replacement of the membrane and electrolyte).

3.1.3 *Dissolved oxygen meters*

Fermenter and probe manufacturers offer instruments to which dissolved oxygen probes of either or both types may be connected, to give calibratable readings of oxygen tension. Note that the electrical interaction is quite different with the two types of sensor, so it will be necessary at least to set a switch depending on whether a galvanic or polarographic probe is in use. It is possible to economize in the area of dissolved oxygen meters, unlike probes. A galvanic probe can be quite successfully used to directly generate a voltage input to a potentiometric chart recorder. The probe leads should be connected through a suitable load resistor, of the order of 500 Ω, and the voltage across this connected to the recorder input: the resistor should be chosen so that the full scale voltage is round 10 mV (i.e. much lower than the cell potential of about 0.7 V;

this is equivalent to saying the load resistor is much smaller than the effective cell internal resistance). A polarographic probe can be operated with a home-made box containing the simple circuit shown in *Figure 3*, which again provides an mV output. The cell can be any convenient commercial battery giving 1.5 V or so: a 'long life' variety is probably best for calibration stability. R_1 and R_2 should be selected depending on the effective internal resistance of the probe at 100% dissolved oxygen. Manufacturers usually quote a full scale current, which should be divided into the operating voltage of about 0.65 V (e.g. for the quoted 0.1 μA of the Ingold probe, the apparent resistance is 6.5 MΩ). R_1 should be chosen as about 1/50 of this value (i.e. about 100 kΩ for the Ingold probe), while R_2 should be about 1/65 of it. The potentiometric chart recorder should be on a 10 mV scale: the full scale signal may be set by either adjusting a sensitivity setting on the recorder, or by making R_2 variable and altering this. Before operation the variable resistor R_1 should be adjusted until the voltage supplied to the probe is the value required (about 0.65 V with the usual Ag/AgCl reference electrode). The platinum electrode should be negative. For more precise setting, or with a probe whose details are uncertain, study the effect of the voltage supplied on the current passed with the probe in air-saturated medium. The graph should display a plateau, as shown in *Figure 4*, and the voltage should be set in the middle of this. Note however that if you use a probe that provides for temperature compensation, these 'home-made' meters will not be able to take advantage of this feature.

3.2 **Probe preparation and mounting**

3.2.1 *Membranes*

The gas-permeable membrane will need replacing from time to time as it becomes damaged, weakened, or fouled. These defects will show up as drifting or noisy response, or poor response time. In our hands, the membrane modules on the Ingold probe last for several months before needing replacement. With some other probes it seems wise to replace the membrane routinely before each run, to minimize the risk of complete failure during the experiment. Most probes use a teflon membrane of $10-40$ μm thickness (the Ingold probe is an exception in that it also has a thicker silicone membrane: both are fixed inside a module replaced as a whole). With many probes it is possible to provide one's own teflon membrane material, rather than be forced to use that supplied by the manufacturer.

Using a thicker membrane will reduce the risk of damage during sterilization or use, at the expense of a slower response time. When placing the membrane on the probe, make sure that no air bubbles are trapped between the membrane and the probe body or electrode. Great care should be taken in making sure that the membrane is properly sealed to the probe body—failures here are often the cause of problems. This is probably an important area of difference between commercial designs of probe—I suspect it may be impossible to obtain a sufficient seal using a simple rubber O-ring under tension, as used in some makes. *Figure 5* shows two of the membrane fixing systems used. Breakdowns in the seal may cause problems either because broth enters the measuring cell, giving erroneous readings, or because the electrolyte enters the broth and affects the cultured cells (particularly a problem with lead-containing electrolytes for galvanic probes). The membrane is of course very sensitive to mechanical damage: even a pin-

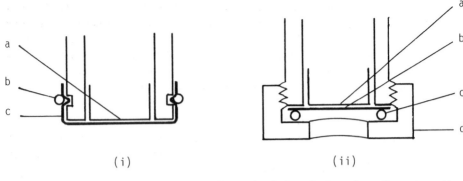

(i) (ii)

Figure 5. Systems for oxygen probe membrane fixing using O-rings. Designs of type (i) may be unable to give the necessary degree of sealing. Type (ii) is usually better, but may give problems if the membrane tends to crumple while tightening the screw-cap; this may cause leaks. (i-a) noble metal electrode; (i-b) rubber O-ring; (i-c) membrane; (ii-a) noble metal electrode; (ii-b) membrane; (ii-c) rubber O-ring; (ii-d) screw cap.

point hole will prevent the probe operating correctly. Hence care should always be taken to prevent it touching anything else while handling the probe, especially while inserting or removing it through the vessel head plate.

3.2.2 *Electrolyte*

The internal electrolyte should normally be replaced at the same time as the membrane (perhaps more frequently with the Ingold probe). As with pH probes, manufacturers supply their own formulated electrolyte solutions, though some laboratories prepare their own. Lee and Tsao (5) list some possible formulations, while I understand the following has been used successfully with galvanic probes: dissolve 202.5 g sodium acetate and 113.7 g lead acetate in 1.5 litre water, add 855 ml acetic acid, and make up to 3 litres with water. In general, the lead electrode of a galvanic probe should be kept completely covered with electrolyte at all times. Unfortunately, some designs do not allow this, and in these cases, the most common cause of final complete failure is breakage of the lead electrode at the electrolyte surface inside the probe: the metal seems to dissolve particularly rapidly at this point. Hence lifetime may be prolonged by periodically changing the electrolyte level.

3.2.3 *Mounting and sterilization*

The position of the probe tip in the broth should be considered carefully. In some fermentations there may be problems with microbial growth on the membrane surface, which will of course lead to an erroneous reading. However, with teflon or silicone membranes I have found this is not usually a problem if the probe tip is in a region of high liquid flow velocity, such as in the side stream from the impeller in a stirred fermenter. However, it can be significant if the probe is in a relatively dead zone for flow, particularly with an organism prone to surface growth, like many filamentous fungi. The probe should be inserted and sealed in to the fermenter before sterilization. Depending on the type of connector used, some protection of the electrical leads may be sensible during autoclaving. The leads of a galvanic probe should be shorted together during sterilization, as this aids the removal of oxygen from the probe interior (sometimes referred to as 'depolarization'), which would otherwise require several hours of stabilization before proper readings are obtained. Some workers use ethanol sterilization of probes, as described for pH sensors in Section 2.2.2 above, but as discussed there, I would not recommend it. If it is used with dissolved oxygen probes, they will need a period of equilibration in the uninoculated fermentation broth. Calibration should not be attempted until the signal from the probe has stabilized.

3.3 Use of the dissolved oxygen probe

When using a dissolved oxygen probe it is important to bear in mind the effect on its reading of three physical parameters – stirring, temperature and pressure. Accordingly, these effects will be discussed first.

3.3.1 *Effect of stirring conditions*

Unlike the pH probe, which makes measurements at negligible current, both types of oxygen probe pass appreciable currents during operation, so significant amounts of oxygen are consumed by the probe. The probe signal is proportional to the rate at which oxygen arrives at the electrode surface, which will in turn be controlled by the rate at which oxygen diffuses across the membrane. This rate will be proportional to the concentration in the bulk of the broth, but the constant of proportionality (and hence the probe calibration) will depend on the characteristics of the overall mass transfer process. It is usual to operate the probe under conditions in which transfer to the outer face of the membrane is very fast and not limiting. Hence the overall process is limited by transfer through the membrane, for which the constant of proportionality (mass transfer coefficient) is more likely to remain constant. Rapid transfer outside the membrane is satisfactorily achieved under any stirring conditions likely to be used during a fermentation experiment, but it is important to realize that the vessel must be stirred even during the initial calibration of the probe.

3.3.2 *Effect of temperature*

Oxygen probe signals increase significantly with increase in temperature, principally because of influences on the rates of oxygen diffusion. In a fermenter it is of course usual to control temperature during an experiment. However, even variations of $0.5°$ or so, which may often persist, can cause significant (more than 1%) changes in probe signal. Cycling of the dissolved oxygen reading (with a period of the order of a minute) is usually a sign of this effect. Furthermore, larger temperature changes can cause major shifts in calibration. This must be remembered if controlled temperature is deliberately changed during an experiment. Also it is essential to calibrate before a run with the fermenter controlled at its operating temperature.

Because of these effects some probes contain integral temperature sensors and with appropriate meters these can be used to give automatic temperature compensation. Alternatively, such compensation can be introduced by software in a computer-monitored or controlled fermenter, using the signal from the separate temperature probe.

3.3.3 *Effect of pressure*

Pressure changes will affect the reading of an oxygen probe, though this in fact reflects a real change in dissolved oxygen. The probe response is essentially determined by the partial pressure of oxygen in equilibrium with the solution around it. The reading is, however, usually expressed as percentage saturation with air at atmospheric pressure (so that 100% 'dissolved oxygen tension' (DOT) means a partial pressure of about 160 mm Hg). However, if the total pressure of the gas equilibrating with the broth varies, the probe reading will change, even if the gas composition is unaltered (because the partial pressure of oxygen will change proportionately). Mathematically, if equilibrium is reached, the probe signal will be determined by:

$$P(O_2) = C(O_2) \times P_T$$

Where $P(O_2)$ is the partial pressure of dissolved oxygen sensed by the probe, $C(O_2)$ is the volume or mole fraction of oxygen in the gas phase, and P_T is the total pressure.

Thus changes in the pressure of gas bubbles in the broth will affect the dissolved oxygen tension and hence the probe reading. In a laboratory-scale fermenter, hydrostatic pressure will not significantly affect gas bubble pressure, but changes in head pressure can do so. Its effects can be significant even when intentional variations are not made — it is common for pressure drops across an exit filter and lines to give a head pressure of 0.07 bar or so, which would be enough to give a 7% increase in probe signal. Furthermore, during the course of an experiment, variations in atmospheric pressure may cause changes: a range of 5% is observed even under normal weather conditions.

Allowing for pressure effects essentially involves making a policy decision about calibration of the probe.

(i)   The probe may be calibrated at atmospheric pressure. In this case DOT values of greater than 100% may be obtained during the experiment. This does not mean that the broth is supersaturated with respect to air of atmospheric composition, merely that the increased pressure of air supplied has increased oxygen partial pressure above that used for calibration.

(ii)   The probe may be calibrated at the expected operating pressure. In this case the 100% reading will represent broth supersaturated with respect to atmospheric air.

(iii)   Perhaps the best, though rarely used approach, is to present all results in terms of 'dissolved oxygen partial pressure', or perhaps 'dissolved oxygen activity', calibrating on the basis of calculated values for the conditions of calibration. This is the parameter that most directly affects probe response.

3.3.4 *Calibration*

The oxygen probe will need calibrating before the fermenter is inoculated at the start of each experiment. A linear calibration is normally assumed, with adjustable zero and slope.

The zero point may be set after extensive gassing of the fermenter with oxygen-free nitrogen. This is best done immediately after sterilization, as this process will have already removed much of the dissolved gases. However, with most probes the output at zero oxygen is very close to electrical zero, so proper zero-point calibration may be dispensed with unless readings are planned at very low dissolved oxygen tensions: instead, one of the probe leads may simply be disconnected to set the current equal to zero. If it is desired to check the calibration zero *after* an experiment, a convenient method is to add a little sodium dithionite to the fermenter contents: this rapidly reacts chemically with any oxygen present. (However, it will also kill the cells, so can only be used afterwards!)

The other calibration may be variously described as slope, sensitivity, full scale, range, etc. It should be set before inoculation, after extensive gassing of the fermenter with air, with stirring and at the temperature of operation. This should give saturation with atmospheric air, and the meter may be set to read 100% dissolved oxygen tension, or the appropriate calculated value of partial pressure. The options concerning the relationship between pressure and calibration are discussed in Section 3.3.3 above.

As with the pH probe, it will not normally be possible to reset the calibration later in the run, so a mechanical locking of the appropriate knob is advisable, to prevent accidental changes.

The comments made on withdrawal of a pH probe for recalibration apply equally to a dissolved oxygen sensor: again devices are available allowing aseptic recalibration, but only for larger fermenters. However, I know of no alternative method for checking the calibration. Hence the importance of selecting a reliable sensor: in critical experiments, using two or more probes in the fermenter may be worth considering, especially if long runs are contemplated.

3.3.5 *Response time*

The response time (95%) of a properly working dissolved oxygen probe should be about 60 s, depending on the thickness of membrane used. This is sufficient for routine monitoring, but is sufficiently slow to cause problems when transient measurements are desired: it has to be corrected for when attempting to measure oxygen uptake or transfer rates by dynamic methods; and is one of the factors contributing to the difficulty of operating an effective control loop for dissolved oxygen tension in a fermenter. The response time is often rather longer for a downward shift in dissolved oxygen compared with an upward one.

Lengthening response time is one sign of deteriorating probe performance with age. Lee and Tsao (5) describe possible methods of recovering an ageing probe by cleaning the electrodes, though I have no experience with these.

3.3.6 *Light effect*

One interference with a polarographic probe can be baffling to trace! Direct sunlight falling on the Ag/AgCl reference electrode can cause a small shift in current. The resulting fluctuations in response as people walk through the light beam can be puzzling. This can of course only be a problem with glass fermenters and probes.

4. REDOX POTENTIAL SENSORS

Heat-sterilizable redox electrodes are commonly offered for fermenters. The first problem with redox probes is deciding what, if anything, their signals tell us about the behaviour of the fermentation process. In this respect a little theory may be helpful.

4.1 **What do they measure?**

4.1.1 *Basic theory of redox equilibria*

The redox probe reports the electrochemical potential generated at a platinum electrode, relative to a reference electrode. This potential will reflect the concentration ratio for the chemical species able to undergo half-cell reactions at the platinum surface. Each of these reactions will involve the interconversion of an oxidizing and a reducing agent: the pair are often described as a 'redox couple'. For example, the redox couple $H_2/2H^+$ is probably important in many cases. Sometimes the couple may include more than one reactant on one or both sides of the redox equation, as in $4H^+ + O_2/2H_2O$.

Each redox couple is characterized by a standard potential, E_0, at which the concentrations (strictly activities) of the oxidized and reduced species involved will be equal (at equilibrium). The potential can only be measured relative to some other couple, and conventionally the standard hydrogen electrode is taken as the zero point (see Section 4.4 below). If the actual potential at an electrode is more positive than E_0 for the redox couple involved, then the equilibrium concentrations for the oxidized species will be greater than for the reduced ones. (This is the usual sign convention – however, you may still find references to the completely opposite convention!)

Platinum electrodes are often used in the study of redox processes at equilibrium in chemical and biochemical systems. In this case every possible redox couple in the reaction mixture will have the same half-cell potential (by definition, if the system is in redox equilibrium). Usually several of them will react directly with the electrode, so that it reports the overall equilibrium position. If the system is titrated with an oxidizing or reducing agent, then allowed to re-equilibrate, the concentration ratios of all the redox couples will change, so that all now reflect a new equilibrium half-cell potential.

4.1.2 *Redox couples in a fermentation broth*

In a fermentation broth, the picture is unfortunately much more confused. Except possibly at the end of the process (and that probably means at the end of the culture death phase!), the culture is very definitely not at redox equilibrium. Cells obtain energy for maintenance and growth by coupling internal redox reactions to other metabolic processes, such as ATP synthesis. It follows that they must take up from their surroundings components that are not at redox equilibrium, so that their reaction is favourable in free energy terms. So while some redox half-reactions may be in equilibrium with each other, they cannot all be. Under these circumstances the potential measured will reflect those couples that react most rapidly at the platinum surface – though quite possibly it will not be exactly in equilibrium with any of them. So the probe signal tells us something about the relative concentrations of uncertain and probably varying chemical species!

If the culture contains even traces of oxygen, then this probably dominates the signal. Oxygen reacts reasonably well at the platinum electrode, and the oxygen/water couple will normally be much more strongly oxidizing than any other present in the medium. Indeed, the redox probe has been suggested for the measurement of trace levels of dissolved oxygen, below the range of the dissolved oxygen probe, and it may have some value here. However, it cannot be safely assumed that no other species will affect the reading. In a more qualitative sense, the redox probe is probably useful for the confirmation of strict anaerobic conditions, by a potential less than about −200 mV. In this case, it is fulfilling a role similar to that of 'anaerobic indicator' dyes often used in bench work.

If the culture is completely anaerobic, the actual signal may respond to a variety of redox couples. In many cases $H_2/2H^+$ may be the dominant couple. Hydrogen reacts well at a platinum electrode (as used in the absolute standard of redox potential) and significant concentrations are common in many anaerobic cultures. Dissolved hydrogen levels may be a useful diagnostic of the state of such fermentations. However, they might be better measured by a technique less liable to interference. For the rich, membrane inlet mass spectrometry can do this on a continuous basis (as well as measure many other dissolved species).

4.2 The equipment

The usual redox potential sensor is a combination probe, including a reference electrode. The indicator electrode is a platinum wire or ring directly exposed to the culture, while the reference will be identical to that used for a pH probe: an Ag/AgCl or calomel electrode in its own electrolyte, connected to the culture through a porous plug liquid junction. Again, special steam sterilizable probes are offered for fermentation work, and some suppliers are noted in Section 6. Because they share essentially the same reference electrode system, many of the considerations noted above for pH sensors apply equally to redox probes. For example, different designs may be needed for autoclave or *in-situ* sterilization (Section 2.1.1).

The electrical behaviour of the sensor is very similar to that of the pH probe, so similar considerations apply concerning electrical connections and amplifier selection (Section 2.1.2). Most pH amplifiers can be simply switched to read out the input voltage directly, as desired for a redox probe.

4.3 Routine handling, sterilization and use

Between fermentations the probe should be well rinsed and stored in distilled water. If cleaning becomes essential due to build-up of organic material on the platinum surface, this may be cleaned by dipping into concentrated nitric acid.

As might be expected, much of what is said above about pH sensors applies equally to redox probes: this includes sterilization (2.2.2), electrolyte (2.3.1), cleaning of the porous plug (2.4.2) and electrical interference (2.4.3).

4.4 Testing or calibration

Unlike pH probes, the zero point and slope of a redox probe does not normally change with time. Thus once the meter calibration is set, it is only necessary to check it from time to time. The meter should read directly the potential of the electrical cell within the probe, variations in which are due to redox reactions at the platinum electrode, since the reference electrode is unchanged.

However, crude potential readings in a given solution will differ from probe to probe, if they use different reference electrodes and/or electrolytes. The difference will consist of a constant offset for all test solutions. By convention, absolute redox potentials are expressed relative to the standard hydrogen electrode. Hence actual probe voltages must be corrected for the potential of the reference electrode used on this absolute scale, either electronically (the zero adjustment on the meter may sometimes be used) or by simple addition. Without correction the voltage readings are at best confusing, if not meaningless, but some workers appear to overlook this point.

The necessary correction may sometimes be supplied by the probe manufacturer. Alternatively, it may be obtained from tables of redox potentials if the nature of the reference electrode is known (e.g. an Ag/AgCl electrode in 3 M KCl is 244 mV more positive than the standard hydrogen electrode, so this value should be added to crude cell potential readings). Finally, the testing procedure described next may be applied to obtain an initial calibration.

The behaviour of the probe may be tested as follows. Saturate one or more standard pH buffer solutions with solid quinhydrone by stirring with excess at a controlled temperature (usually 25°C). Dip the probe into each solution and record the readings. The absolute potential (i.e. corrected to the standard hydrogen electrode) should be +285 mV at pH 7 and 25°C, increasing by 59.1 mV for each unit pH reduction. If the readings deviate by more than 5 mV, there is some fault in the electrode – possibly an incorrect reference electrolyte has been used.

5. DISSOLVED CARBON DIOXIDE SENSORS

Steam-sterilizable dissolved carbon dioxide sensors have been commercially available for the last five years or so. As far as I am aware, the only manufacturer to date is Ingold (see Section 6). The sensor consists of a pH probe immersed in a bicarbonate buffer solution, separated from the culture by a gas permeable membrane. Carbon dioxide diffuses across the membrane and alters the pH of the buffer, which is detected by the pH measuring system: the pH may be related to the partial pressure of dissolved carbon dioxide in the culture medium.

There is evidence that dissolved carbon dioxide levels can sometimes have an important effect on the physiology of cells in a fermentation culture. However, at present the dissolved carbon dioxide sensor is not yet widely used in fermentation studies, partly because its price remains high.

6. SUPPLIERS

This information is collected here because there are common factors concerning the suppliers of all types of probe covered in this chapter.

Most fermenter suppliers will offer these sensors as agents for the original manufac-

Table 1.

Manufacturer	pH probes	DOT probes (Galvanic)	DOT probes (Polarographic)	Redox probes
Ingold	Y		Y	Y
Kent (was EIL)	Y			
LSL/Biolafitte		Y		
New Brunswick		Y	Y	
Russell	Y			Y
Uniprobe		Y		

The addresses of these suppliers are listed in the appendix.

turer. Often a good financial deal can be had through them, particularly if the sensors are bought as part of a complete fermenter. Presumably there is a substantial retail mark-up whether the probes are sold direct, through a specialist agent, or through a fermenter dealer!

Table 1 gives the original manufacturers of heat-sterilizable fermentation sensors that I know are currently offered for general sale in the UK.

7. FURTHER READING

pH probes

Most of the standard texts on biotechnology, biochemical engineering, or fermentation technology devote a paragraph or two to pH sensors, but I known of no recent attempt to cover probes for fermenters in depth.

The general principles of the use of glass electrodes for pH measurement are covered to some extent by most texts on practical or physical bioscience. Some more detailed references are:

1. Linnet,N. (1970) *pH Measurement in Theory and Practice*. Radiometer, Copenhagen.
2. Westcott,C.G. (1978) *pH Measurements*. Academic Press, New York.
3. Covington,A.K. (ed.) (1979) *Ion Selective Electrode Methodology*, Vol. 1. CRC Press, Boca Raton, Florida.
4. Linder,P.W., Torrington,R.G. and Williams,D.R. (1984) *Analysis Using Glass Electrodes*. Open University Press, Milton Keynes, UK.

DOT probes

5. Lee, Y.H. and Tsao,G.T. (1979) *Adv. Biochem. Eng.*, **13**, 35.

Redox probes

6. Kjaergaard,L. (1977) *Adv. Biochem. Eng.*, **7**, 131.

CHAPTER 7

Instrumentation and con

PETER J.B.DUSSELJEE and JAN FEIJEN

1. INTRODUCTION

Laboratory fermenters are used for fermentation process research and for fundamental microbiological research. The main goal of this research is to acquire the characteristics of a given fermentation process and/or microorganism. These characteristics are derived from monitored process variables and biochemical and physicochemical analyses. The research objective will determine the essential process variables to be monitored and/or controlled and the necessary analytical measurements to be performed.

These introductory remarks implicity determine the demands for instrumentation and control of a laboratory fermenter. This chapter is concerned with the elaboration and the practical realization of these demands.

1.1 Premises and definitions

Just as biotechnology is the integration of several scientific disciplines, even more is instrumentation and control of (laboratory) fermenters a field in which semantics and jargon can often cause dramatic misunderstandings ('unpublished results'). A glossary of terms is therefore given at the end of the chapter, and some concepts are explained.

The laboratory fermenter is a bioreator located in a laboratory environment and employed to carry out experiments for the acquisition of information. Consequently, there is a demand for flexibility (i.e. suitable for a variety of products, microorganisms, and operational procedures). In addition, skilled laboratory staff carry out the experiments.

It is often stated that the difference between laboratory and pilot-plant equipment is the size of the equipment. However in some industries small-scale fermenters (25−50 litres) are part of their pilot-plant facilities. Strictly speaking the objectives of a pilot plant and the involvement of less skilled personnel effect requirements for instrumentation and control more than the size of the equipment. As a result of this, pilot-plant facilities are often 'automated' (sequence control; e.g. procedures for cleaning in place sterilization procedures, etc.) in order to improve the control of the processes and the efficiency of operation by reducing the manual operation. Sequence control for laboratory fermenters is outside the scope of this chapter, since it is only occasionally essential for a research experiment.

The bioreactor, including the sensors, the transmitters, the actuators, and the transducers, is referred to as *fermentation equipment*. The term *fermentation system* refers to the application of control engineering and computer technology for the control of the experiment, of the process, and of the process variables. The hardware

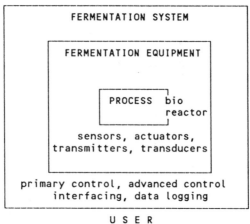

Figure 1. Set-up of a fermentation unit.

configuration and the performance of the fermentation system depend on the demands of the user. These demands include aspects of quality, expansion of experimental possibilities and the exchange of manual operation (e.g. clerical duties) for software and hardware of the fermentation system. The *fermentation unit* comprises both the fermentation equipment and the fermentation system. In *Figure 1* the set-up of a fermentation unit is shown.

1.2 Scope of this chapter

Many publications covering aspects of instrumentation and control of fermentation processes have appeared (1−6), in particular the application of computer technology. It is not the intention of this chapter to review the entire field of the subject. We have confined ourselves to a basic and practical approach for users of laboratory fermenters.

This chapter is structured with sections on fermentation equipment, fermentation systems and the description of representative fermentation units.

2. FERMENTATION EQUIPMENT

The fermentation equipment covers a wide range of bioreactor designs (see Chapter 2). Regardless of the design of the bioreactor, the variables given in *Table 1* could be measured for a bioprocess. A sensor is required for each measured variable. Each sensor needs a transmitter for the modification of the information from the sensor into a well-defined output signal. The output signals are connected to the fermentation system. The sensor and the transmitter are sometimes mounted together in one measuring device.

2.1 Sensors

In general a sensor can be characterized by the indication of its operation in relation to its application for process control.

(i) *Off-line sensor.* This sensor is not part of the fermentation equipment. The measured

Table 1. Process variables.

Physical variables	Chemical variables	Biological variables
Air (gas) flow	Dissolved carbon dioxide	Active biomass
Bioreactor volume	Dissolved oxygen	Extracellular biochemicals and
Conductivity	Fluorescence	metabolites
Foaming	Ionic strength	Intracellular components:
		DNA
Liquid flow	Nutrient concentrations	RNA
		NADH
Optical density	Off-gas constituents	ATP
		proteins
Power input	Osmolarity	amino acids
(agitator speed)		Microbial population
Pressure	pH	(contamination)
	Redox potential	
Temperature		Morphology
Viscosity		Total biomass

value is not directly available for process control. The intervention of an operator is essential, first for the actual measurement and finally for entering the measured value into the fermentation system for process control, if required. Despite much research effort (7–8), for many of the process variables in *Table 1* only off-line sensors are available for routine use.

(ii) *On-line sensor.* This sensor forms part of the fermentation equipment, but the measured value of the variable is not directly available for process control. The intervention of an operator to enter the measured value into the fermentation system for process control is necessary.

(iii) *In-line sensor.* This sensor is part of the fermentation equipment and the measured value is, without intervention of an operator, directly used for process control.

Since the difference between on-line and in-line is only a result of the application of the sensor for process control, in this section only the indication off-line or on-line will be used for the sensor.

The output signal of a sensor is also an aspect of concern. By its mode of operation all sampling routines (even on-line analytical instruments like HPLC or a mass spectrometer) produce a discontinuous output signal.

2.2.1 *Sensors monitoring physical variables*

(i) *Air (gas) flow.* Thermal mass flow sensors are reliable and accurate sensors for gas-flow monitoring, virtually independent of pressure and temperature changes. The principle of this sensor is based on a heat balance. Two temperature sensors are elements of an electrical bridge circuit measuring the temperature before and after a heated section.

151

The thermal mass flow sensor is an on-line sensor, which produces a continuous, analogue output signal.

(ii) *Bioreactor broth volume.* The bioreactor broth volume is best measured as the weight of the fermentation broth. An analytical balance based on the principle of electromagnetic force compensation for measuring total bioreactor weight needs special attention during operation. Tubing, wiring, etc. fitted to the bioreator can cause anomalous readings, but this sensor has proved to be applicable for routine use.

The analytical balance is an on-line sensor, which produces a continuous, binary coded decimal (BCD) output.

An alternative for an analytical balance is a load cell. The total bioreactor weight causes micro-deformations in the cell which are measured by a piezoresistive element which is part of a Wheatstone bridge circuit.

The load cell is an on-line sensor, which produces a continuous, analogue output signal.

(iii) *Conductivity.* Conductivity of the fermentation broth can be monitored by a simple cell consisting of an anode and a cathode. Additional electronic instrumentation (frequency oscillator and a Wheatstone bridge) is required for the actual measurement of the impedance.

This electrode is an on-line sensor, which produces a continuous, analogue output signal.

(iv) *Foaming.* Two simple and reliable conductance probes (insulated stainless steel wire) form an electrical circuit. One probe is immersed into the fermentation broth or mounted onto a metal part of the bioreactor, which is in contact with the fermentation broth. The other detects the level of foam in the bioreactor.

This electrode is an on-line sensor, which produces a continuous, discrete output signal.

(v) *Liquid flow.* By weighing the feed tank the liquid flow can (indirectly) be monitored. The analytical balance is an on-line sensor, which produces a continuous, binary coded decimal (BCD) output.

Alternative flowmeters (e.g. magnetic flowmeters, mass flow meters based on the principle of modulated vibrations in a U-tube, turbine flowmeters) have special claims for sterilization (e.g. *in-situ* steam sterilization).

(vi) *Optical density.* The sensor for optical density (turbidity) can be any commercial off-line spectrophotometer or a sterilizable optical cell system (Eur-Control). The latter is an on-line sensor, which produces a continuous, analogue output signal.

(vii) *Power input.* For a laboratory bioreactor the stirrer speed is used as an indication for power input. For on-line measurement of the stirrer speed a tachogenerator (dynamo) on the motor shaft will produce a continuous, analogue output signal proportional to the stirrer speed.

(viii) *Pressure.* Any commercially available pressure gauge covered with a membrane (diaphragm) can be used. The pressure-sensitive element (membrane) is mounted directly to a strain-gauge or a piezoresistive element. The strain gauge or piezoresistive element is part of a Wheatstone bridge configuration.

These pressure-measuring devices are on-line sensors, which produce a continuous, analogue output signal.

(ix) *Temperature*. The recommended sensor for temperature measurement is the thermoresistance element Pt100. The platinium resistance is calibrated at 0°C to 100 Ω. The electrical resistance varies with a change in temperature. The resistance is mounted in a protection tube.

The Pt100 element is an on-line temperature sensor, which produces a continuous, analogue output signal.⁄

(x) *Viscosity*. Viscosity of fermentation broth is better measured off-line in one of the commerically available viscometers producing a discontinuous measured value. Impeller viscometers have proved to be advantageous for microbial suspensions.

Present on-line rheometry (9 – 10) is too limited for routine use in laboratory fermentation equipment, since rheological characterization needs a range of shear-rate adjustments.

2.1.2 *Sensors monitoring chemical and biochemical variables*

Since in Chapter 6 and Chapter 10 principles and operation of the measurement of the chemical and biochemical variables are elaborated, only an overview in *Table 2* is given for the sensors involved.

2.2 Transmitters

Besides the actual sensing device, specific electrical circuits (Wheatstone bridge, amplifier, etc.), of which the sensing device is often part, are also applied. Then the output signal is conditioned for coupling to the fermentation system. This instrumentation is referred to as a transmitter. We shall consider transmitters as black boxes.

In order to ease flexible coupling of the fermentation equipment to a fermentation

Table 2. Sensors monitoring chemical and biochemical variables.

Variable	Sensor	Output signal
Dissolved carbon dioxide	Electrode	Continuous, analogue
Dissolved oxygen	Electrode	Continuous, analogue
Fluorescence	Optical cell	Continuous, analogue
Nutrient concentrations	Ion-specific electrode	Continuous, analogue
	(Instrumental) assay	Discontinuous, BCD
Off-gas constituents	Instrumental assay	Continuous, analogue[a]
Osmolarity	Instrumental assay	Discontinuous, BCD
pH	Electrode	Continuous, analogue
Redox potential	Electrode	Continuous, analogue
Active biomass[b]	Assay	Discontinuous, BCD
Extracellular biochemicals and metabolites	(Instrumental) assay	Discontinuous, BCD
	Specific biosensor	Continuous, analogue
Total biomass[c]	Assay	Discontinuous, BCD

[a]When shared with other fermentation units: discontinuous as result of sampling system.
[b]Indirectly estimated via fluorescence.
[c]Indirectly estimated via optical density.

system the application of standard output signals is recommended:

(i)   *continuous, analogue and discrete output signals*: $0-10$ V (direct current) or $0/4-20$ mA (direct current), and to avoid problems with grounding the signal should be floating.

(ii)   *binary coded decimal output* should be transmitted by standard RS232, RS423, or IEEE488 interfaces and their communication protocols.

2.3 Actuators and transducers

The actuator is the element of the fermentation equipment, which alters the value of a monitored variable in response to the output signal of a controller or manual intervention of the operator. Again, standard signals are recommended (see section 2.2). These signals are modified into an applicable signal for the actuator by a transducer, if required. The actuator acts continuously (e.g. control valve, speed of motor or pumps) or discontinuously (e.g. values open/closed, pumps on/off).

Most actuators are described in Chapter 1 (pumps, etc) and Chapter 2 (agitation system, sampling systems).

3. FERMENTATION SYSTEMS

All functions for the control of a fermentation experiment, the control of the resulting fermentation process and the control of the corresponding process variables are performed by the fermentation system, except the manually operated functions. In general the following functions can be itemized:

(i)   The first task of the fermentation system is to enable connecting of the process signals to the fermentation system: *process interfacing*.

(ii)   The next task of the fermentation system is to control the selected process variables: *primary control*.

(iii)   Another task might be to control the actual process: *advanced control*. Advanced control includes setpoint generation, supervision, alarming, and multivariable control.

(iv)   The next task is to monitor the (off- and on-line) measured process variables: *data logging*.

(v)   Another task is to display various data to the operator and to facilitate manual input: *operator interfacing*.

(vi)   The researcher requires the collection, processing, and storage of data logged from the experiment: *researcher interfacing*.

(vii)   Finally, the operation of the fermentation experiment can be 'automated' (i.e. replacing manually operated tasks): *sequential control* (e.g. medium preparation, sterilization, harvesting, cleaning).

Formerly, the functions of the laboratory fermentation system were limited to monitoring and controlling a few process variables, including process and operator/researcher interfacing. As stated before (section 1.1) it is the user who specifies the functions to be performed by the fermentation system ('market pull'), with the possibility of a flexible implementation of functions to be performed by the fermentation system, replacing manual operations.

In the following sections some of above functions are discussed in more detail.

3.1 Process interfacing

The input signals for the fermentation system are the output signals from the fermentation equipment (see section 2.2).

For (conventional) electronic controllers the continuous, analogue input signal can be directly coupled to the controller system.

As soon as computer technology (e.g. microprocessor-based controllers) is involved, input signals have to be converted into a digital equivalent, and analogue output signals have to be generated from a digital equivalent. For continuous, analogue signals ADCs and DACs are used (ADC: analogue-to-digital converter; DAC: digital-to-analogue converter). To meet an accuracy of 1 % for the conversions, the digital equivalent should comprise at least 10 bits. For continuous, discrete signals DIs and DOs are used (DI: digital input; DO: digital output). This digital signal comprises one bit. A DO signal should preferably come from a relay.

For a discontinuous, analogue input signal (e.g. shared instrumental sensor, result of sampling procedures) both an ADC and a DI are needed. The discrete signal acts as a flag for a quality indication of the analogue signal.

Data interfaces are required for the coupling of devices using processors for monitoring or controlling variables. The information transferred are BCD numeric values. RS232 and RS423 are standard serial transmission interfaces and are widely used. However, this standard is sometimes 'firmware' adjusted! The disadvantage of this type of interface is its relatively slow transmission rate.

Some analytical instruments are equipped with a byte-oriented, parallel transmission interface: IEEE488 (formerly known as an HPIB interface). The disadvantage of this type of interface is the limited distance for transmission (15 m).

3.2 Primary control

Primary control in its simplest form is the open-loop control. The actual value of the variable to be controlled is not used for adjustment of the setpoint. The setpoint is derived from a calibration procedure beforehand, e.g. nutrient flow by adjustment of the speed of the pump, without any flow-metering device.

As soon as any measured value of the variable to be controlled is used for (re-)adjustment of the setpoint, there is closed-loop control. Depending on the intervention of the operator for closed-loop control, the sensor can be off-line, on-line or in-line (see section 2.1). For example, nutrient flow is measured by a flow-metering device; the readout of that device is used for manual (off-line or on-line sensors) or self-adjustment (in-line sensors) of the setpoint to the value required.

An elementary controller for closed-loop control is a stand-alone analogue controller (*Figure 2*). Both the input and the output of this controller are analogue signals. Primary control of a fermentation process in principle requires several of these controllers.

There is a trend to replace this type of controller by a (micro)processor-based stand-alone controller. The advantage is the potential for dedicated (model-based) and advanced control algorithms and cost effectiveness (e.g. purchase, fail safe).

3.3 Advanced control

The impact of the recent developments in computer technology in the field of control

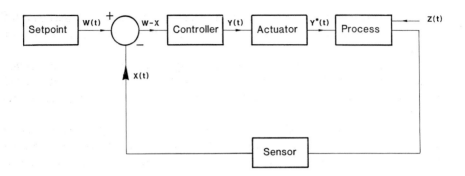

Figure 2. A diagram of a stand-alone analogue control loop.

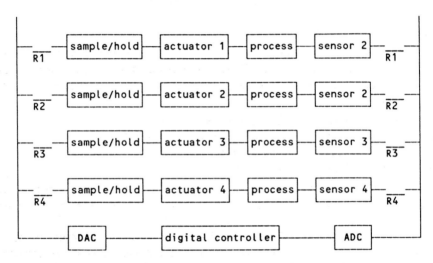

Figure 3. A diagram of a DDC system containing several control loops.

engineering is considerable. Sometimes advanced control is thought to be synonymous with the application of computer technology. (Micro)computers can be integrated with the primary controllers. Similarly to the integration of sensors for control, the (micro)computer can be integrated off-line, on-line, and in-line.

The in-line application of (micro)computers also offers the opportunity, to replace the primary controllers by direct digital control (DDC) systems. The remaining in-line applications of (micro)computers are the hierarchical control systems (indirect digital control).

3.3.1 *Direct digital control*

Direct digital control is mostly applied in processes which contain a large number of control loops. In *Figure 3* a diagram of a DDC system is given.

In the basic DDC system the various loops are controlled by only one type of controller. This controller controls the loops in succession.

When the relay contacts R_i are closed the relevant loop is connected to the controller and the controller controls the variable. Both the relays and the control parameters are controlled by the system.

A drawback of direct control is that when the digital control system breaks down, the entire process is without control. A second disadvantage is that in the basic DDC systems all loops have to apply the same control algorithm and only the control parameters can be adjusted.

The former advantage of DDC was its economic attractiveness for processes containing many control loops. However, in recent years the cost of controllers has become a minor part of the total investments for a fermentation system, so this consideration is now of little importance.

Besides the functions of primary control, most DDC systems have functions for advanced control, operator, and researcher interfacing.

3.3.2 *Hierarchical control*

Hierarchical control is the in-line integration of a (micro)computer with primary controllers without replacing them. Consequently at least two levels of control of the process variables can be distinguished in hierarchical control systems. Often more levels are defined:

Level 0 control: process interfacing (see Section 3.1).
Level 1 control: single-loop controllers for primary control.
Level 2/3 control: multivariable controllers.
Level 4 control: expert system control.

In addition to these functions related to the control of the process variables, the fermentation system will also perform other functions (e.g. operator and researcher interfacing, setpoint generation, supervisory control).

Level 2 and 3 controllers adjust the setpoint of level 1 controllers or the actuator(s) of level 1 controllers. The difference between level 2 and level 3 controllers is the time base for the control action. Typical values for the time interval of level 2 controllers are between 1 s and 1 min. The time interval of level 3 controllers can vary from minutes to several hours. Both these controllers are also referred to as master – slave controllers. An example of this is a dissolved oxygen concentration controller which uses the stirrer speed controller and the pressure controller as slave controllers. Typical examples of level 3 controllers are control loops which control process variables which are determined by means of off-line analysis (e.g. nutrient concentration).

Level 4 control needs a certain expertise (expert system).The controller itself is comparable to a level 3 controller. However, for a level 4 control action an evaluation with a degree of uncertainty is essential. Techniques such as pattern recognition (a comparison with historic experimental data) or modelling are used for an evaluation of the state of the process and its variables. Since the interpretation and evaluation of several phenomena is necessary, the control actions are mostly presented as advice to the operator (advisory control). Only in a few applications are level 4 control actions performed automatically.

So far only level 1, 2, and 3 control are to be found in commercially available laboratory fermentation systems. For level 4 control an on-line (micro)computer is often

PROCESS CONTROL   CONTROL OF PROCESS VARIABLES

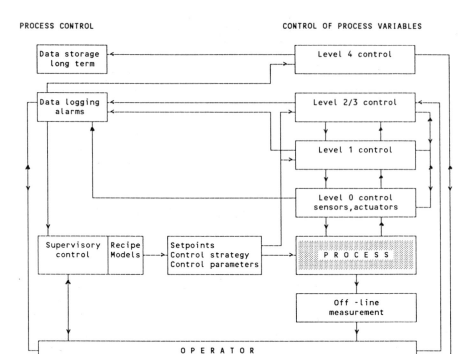

Figure 4. A schematic diagram of a hierarchical control system.

applied. Under these conditions level 4 control can also be referred to as process evaluation or computer-aided research. In *Figure 4* a schematic diagram of hierarchical control is given.

In most hierarchical systems the hierarchy is not only functional but also physical. An example of this is the hierarchical system where advanced control and data storage is located in a (so-called) 'process computer', and the primary control is performed by an 'instrumentation computer'. In these fermentation systems the process interfacing is the responsibility of the instrumentation computer (*Figure 5*).

The disadvantage of process control being shared by two different computers is the need for communication between the two computers. Much data has to be transported from the instrumentation computer to the process computer and vice versa. In applications where one process computer is shared by several instrumentation computers, this can cause communication problems.

In modern control systems the various functions are distributed among many processors. A future development for fermentation systems is possibly, the distributed hierarchical system (*Figure 6*). Here the information of the process(es) is directly available to all the modules, each module having its own processor and data storage. The advantage of these systems is that data transfer between the modules is limited to data which are essential for a specific module. In these distributed systems the data transfer rate is very high. Another advantage of distributed systems is that redundancy of both data transfer and control modules are more or less standard features of the system.

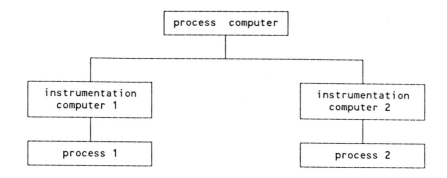

Figure 5. A schematic diagram of heirarchical control system where the process control functions are performed by an instrumentation computer and a process computer.

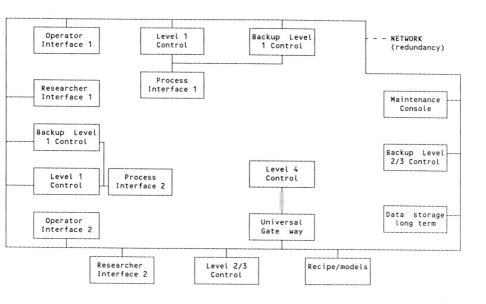

Figure 6. A diagram of distributed hierarchical fermentation system.

3.4 Control engineering

In simple terms control can be defined as the process in which a measured physical quantity (x) is compared to a setpoint (w), and the deviation ($w-x$) is reduced or suppressed by means of a prescibed control algorithm (*Figure 2*).

The application to single-loop controllers is straightforward. However, the application to level 2 and 3 controllers is somewhat more complicated. These controllers can use the input signals of more than one sensor, and can produce various outputs.

For single-loop controllers three types of control loop can be distinguished: feedforward, feedback, and adaptive control loops. Combinations of these types of

159

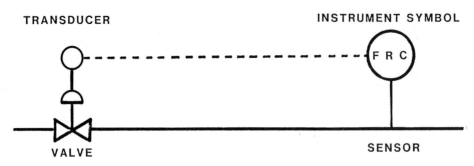

Figure 7. A drawing of a flow controller in a process flow diagram. F, R and C indicating Flow, Recording and Controlling.

control loops are also applied. In the following sections this will be explained in more detail.

In process flow diagrams the symbol used for an instrument is a circle. In this circle a letter code of the measuring function and the readout function is given. The symbol is connected to the process by a solid line (the measuring line) and to the actuator by a dotted line (the actuator line) (*Figure 7*).

The symbol of a transducer is smaller than the symbol of an instrument (usually 5 and 10 mm respectively). When the valve is directly manipulated by the instrument the transducer symbol is not used.

Some common measuring functions and their letter designators are as follows (see also the glossary and *Table 4*):

F	flow	P	pressure
L	level	T	temperature

The readout functions are:

I	indicating	C	controlling
R	recording	A	alarming

3.4.1 *Feedback control*

Feedback controllers compare the measured value of the process variable that needs to be controlled with its setpoint and adjust an actuator in order to suppress the deviation between the measuring value and the setpoint (*Figure 2*). Two ways of feedback control can be distinguished: on/off feedback control and modulated feedback control.

The response of an on/off temperature controller in an endothermic process is given in *Figure 8*.\When the measuring value exceeds the setpoint the actuator output ($Y^*(t)$) is zero. When the measuring value is lower than the setpoint, the actuator output (heating) is 100%. It will be obvious that this kind of controller is unsuitable when the allowable deviation between measuring value and setpoint is small.

Especially in systems with a relatively long response time, on/off control can lead to a serious overshoot. This overshoot can sometimes be diminished when the on-status of the on/off controllers is combined with a pause/pulse setting (*Figure 9*).

In order to achieve this the time delay of the pause-pulse on/off controllers should be set at a value higher than the response time of the loop. It will be clear that when

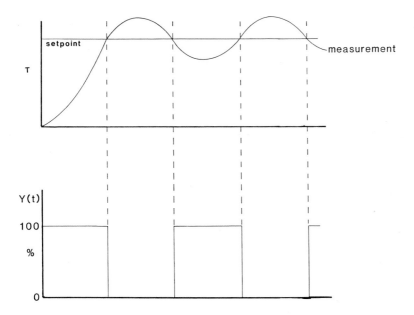

Figure 8. The response of an on/off controller.

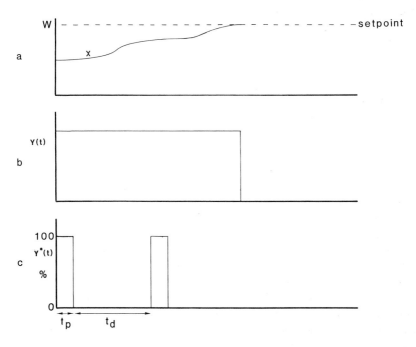

Figure 9. The response of a pause-pulse on/off controller. X is the measured value, W the setpoint, Y the actuator position, t the time, t_p the pulse time, and t_d the delay time. In (b) the controller status is given.

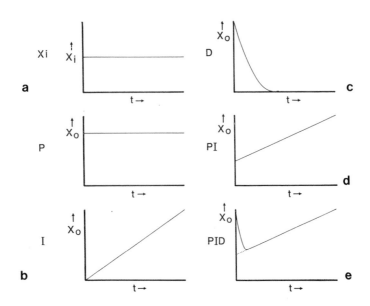

Figure 10. The step responses of P, I, D, PI and PID controllers. x_0 is the output, x_i is the input.

pause-pulse on/off controllers are applied the dynamic behaviour of the control loop is no longer important.

Both on/off control and pause-pulse on/off control are widely applied in fermentation systems (section 3.5).

In processes where the allowable deviation between setpoint and measuring value is small, modulating feedback control should be applied.

Modulating feedback controllers change the actuator position gradually. The dynamic behaviour of these controllers can be adapted to the dynamic behaviour of the control loop.

Well-known examples of modulating feedback controllers are proportional (P) controllers, integrating (I) controllers, differentiating (D) controllers, and controllers that apply a combination of these control modes. Modulating feedback controllers can also be combined with a pause-pulse setting or a time delay. /

In a proportional controller the actuator position is proportional to the deviation between setpoint and measuring value. The response of a proportional controller to a step function is given in *Figure 10a*. The ratio between the actuator position and the deviation between setpoint and measuring value is the proportional factor K_p.

In an integrating controller the actuator position changes at a rate which is proportional to the error (the deviation between setpoint and measurement) (*Figure 10b*). The rate at which the actuator position changes is the integrating factor K_i./

The response of a derivative controller is given in *Figure 10c*. In a derivative controller a sudden change in the deviation between setpoint and measuring value results in a sharp change of the actuator position.

Two well-known combinations of the controllers given above are the PI controller and the PID controller. The step responses of these controllers are given in *Figure 10d*

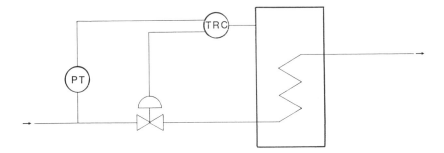

Figure 11. Feedback and feedforward temperature control in a reactor vessel.

and *10e*. The PI controller first changes the actuator position according to the proportional action. After that the actuator is manipulated by the integrating action. In a PID controller the control action is also influenced by the differentiating action.

As will be seen in section 3.5, PI and PID controllers are widely used for fermentation processes.

3.4.2 *Feedforward control*

Feedforward controllers use measured values of variables other than the measured value of the process variable that has to be controlled to carry out a control action. These variables are referred to as disturbance quantities. In *Figure 2* this would mean that the controller uses the value of variable $z(t)$ in order to control the variable $x(t)$. Feedforward control is more often used in combination with feedback control.

An example of a feedforward control action is given in *Figure 11*. The temperature of a reactor vessel is controlled by measuring the temperature of the vessel and manipulating the valve in the cooling water system (feedback control).

The pressure of the cooling water system is measured but not controlled. When the pressure in the cooling water system changes, the cooling water flow will also change. However, as heat transfer is a slow process, this change in the cooling water flow will not lead to an immediate change in the reactor temperature. A simple feedback temperature controller will not adjust the valve position to the required new value until the temperature has changed, thereby allowing the process to be upset. If the change in cooling water flow is sensed as soon as it occurs and is used to correct the valve position, the temperature will not be disturbed. The process perturbation is 'fedforward' and used without waiting for the feedback controller to detect the error in temperature.

3.4.3 *Adaptive control*

In addition to classical feedback and feedforward control, many other modes of control have been developed in the last few years. In literature most of these control modes are referred to as adaptive control. Adaptive control is mostly applied in systems where the static or dynamic behaviour of the process changes with time (non-stationary processes). The adaptive controller can either use on-line measured process data, a theoretical model, or a combination of these two, to predict the change in the static and dynamic behaviour of the process.

As there are many ways to predict a change in a process it will be clear that no general description of adaptive control can be given.

Examples of adaptive control are self-tuning controllers and adaptive pH control. In section 3.5 adaptive pH control will be dealt with further.

3.5 Control in fermentation processes

Because of cell growth, substrate feeding, and product formation the properties of the fermentation broth change with time. This means that fermentation processes exhibit a continuing transient behaviour. However, this transient behaviour is not such that fermentation processes can only be controlled by adaptive controllers. On the contrary, in a lot of fermentation systems only feedback controllers are applied.

In this section the modes of control of the most common fermentation control loops are elaborated. The common modes of control for the various fermentation process variables are given in *Table 3*.

(i) *Temperature control.* In glass laboratory-scale fermenters the temperature can be controlled by an infrared lamp (heating) and cooling coils (cooling). As the response of the system to infrared heating is fast, on/off control is mostly quite accurate. As already explained in section 3.4.1 this might not be the case with on/off cooling. Here, deviations of several tenths of a degree may occur dependent on the cooling equipment. The need for feedforward and modulating feedback temperature control will diminish when the pressure of the cooling water system is nearly constant (no feedforward control needed) and when the cooling capacity of the cooling coils does not exceed the required

Table 3. Control modes in fermentation processes.

Variable	Mode of control
Gas flow	Modulating feedback control (P, PI, or PID)
Stirrer speed	Modulating feedback control (P, PI, or PID)
Pressure	Modulating feedback control (P, PI, or PID)
Temperature	Feedback control (P, PI, PID, or on/off)
	A combination of feedforward and feedback control
	Adaptive control
pH	Feedback control (P, PI, PID, or on/off)
	Adaptive control
Level of foaming	On/off feedback control
Nutrient flow	Feedback control (P, PI, PID, or on/off)
(Flow control)	A combination of feedforward and feedback control
Nutrient flow	Non-standard control which includes modulating
(Weight measurement)	feedback control
Dissolved oxygen concentration	(Multivariable) level 2 control
Off-gas oxygen/carbon dioxide concentration	(Multivariable) level 2 control
Calculated variables/derived quantities/models	(Multivariable) level 2 and level 3 control
Off-line measured variables	(Multivariable) level 3 control

cooling capacity by several factors (little overshoot).

In jacketed laboratory-scale fermenters, both heating and cooling are done by the jacket. Here the same consideration as above is valid.

(ii) *pH control.* In some fermentation processes pH control can be very difficult. The reason for this is twofold. Firstly, the response of a fermentation broth to an acid or base addition can change in the course of the fermentation, and secondly the pH value is a non-linear function of the $[H^+]$ concentration in the broth (pH $= \log [H^+]$).

On/off controllers are not able to adapt to a change in the buffer capacity of the broth. Therefore, they are not suitable for use in fermentation with a considerable change in the buffer capacity. Due to the I-action, PI controllers can do a better job in this respect.

In recent years adaptive pH controllers have also been developed. These controllers use a model and the pH response to previous acid or base additions to adapt its tuning. As already stated in section 3.4, most pH controllers use a 'dead zone' in order to avoid successive acid and base additions.

(iii) *Nutrient flow.* Peristaltic pumps are widely used for feeding liquid to a laboratory bioreactor. The liquid flow then is controlled by an open loop controller: the speed of the pump is manually adjusted to the setpoint of the required liquid flow. By means of a preliminary calibration procedure the liquid flow is known as a function of the speed of the pump. For mass balancing the feed tank is weighed before and after the experiment. However, even for its most simple application, anomalous feed rates can frequently appear. For programmed feeding (e.g. time-dependent profile) sensing the liquid feed rate at adequate time intervals and adjustment of the setpoint are essential.

Based on published principles $(11-12)$ Sartorius developed a flow control element using an analytical balance as flow sensor. The feed tank is (automatically) refilled from a large storage vessel.

The flow can be controlled by, for instance, a PI controller. The problem with this flow control, however, is that the flow is determined by subtracting two discrete values. Depending on the sample frequency, this can lead to unacceptable flow variations. This problem can be solved by using an adaptive control algorithm.

(iv) *Multivariable level 2 and 3 control.* Level 2 and 3 controllers manipulate the setpoint of level 1 controllers to keep the control variable of the level 2 or level 3 controller at the required value. An example of this is a dissolved oxygen concentration controller which changes the setpoint of the stirrer speed controller, the airflow controller or the feed controller to eliminate the error in the dissolved oxygen concentration. Variables as the exhaust oxygen concentration and calculated variables like the OUR can also be controlled by means of such multivariable controllers.

(v) *Models.* Models can be used for controlling both the fermentation process and the process variables. An example of a controller which uses an experimental model is the growth rate controller (13). With this model the growth rate of the organisms is calculated from the carbon dioxide production. Deviations between the calculated growth rate and its setpoint are corrected by adjusting the nutrient flow.

An example of a process control model is the model which is used in adaptive pH control. This model describes the relation between pH response and acid or base addition.

It will be clear that in automation systems which are used for testing control of both fermentation process and process variables a software structure is needed which allows an easy implementation of these models.

3.6 Architecture of fermentation systems

In addition to the functions mentioned earlier in this chapter, other aspects also determine the hardware configuration and the performance of the fermentation system. For fermentations with a high risk (economically or environmentally), failsafe or redundancy capabilities are important. In commercially available laboratory fermentation systems these items are not present.

The architecture of fermentation systems currently in operation is also a result of the historical (evolutionary) development of the system; e.g. original, simple fermentation systems for the control of several process variables often were first expanded by data-logging facilities for the monitored process variables. This expansionary (micro)computer system was often shared by several bioreactors. This aspect of an evolutionary build-up of a fermentation system is often overlooked, especially by suppliers of fermentation equipment. Therefore, a modular-based fermentation unit is highly recommended.

4. FERMENTATION UNIT

The key for a set-up of a laboratory fermentation unit is to decide what experimental information is needed and how to acquire that information. A general approach is the notion of a limiting factor controlling the fermentation process within a defined environment (14). This concept is adopted by the Working Party on Bioreactor Performance of the European Federation of Biotechnology. Based on the publications (15 − 16) of this committee a representative fermentation unit can be defined: the basic laboratory fermentation unit. For simplicity the fermentation equipment is here considered to be an aerated stirred bioreactor for fed-batch operation unless stated differently.

4.1 Basic laboratory fermentation unit

The basic laboratory fermentation unit reflects a minimum level of control for the essential variables (*Figure 12*).

(i) *Off-line monitored variables.* Bioreactor volume, flow (cumulative) of acidic and alkaline titrant, flow (cumulative) of antifoam agent, (optical density), (viscosity), nutrient concentrations (e.g. nitrogen, phosphate, carbon), active/total biomass, extracellular key-biochemical and key-metabolite.

(ii) *On-line monitored variables.* Dissolved oxygen tension, off-gas carbon dioxide concentration, off-gas oxygen concentration.

(iii) *In-line monitored variables.* Air flow, level of foaming, flow nutrient, stirrer speed, temperature, pH. This implies six primary control loops.

Additional demands on the performance of the fermentation system are: manual setpoint feedback control for primary control loops, data-logging with analogue recording instruments (recorder) and off-line data-sheets, data-processing by an off-line

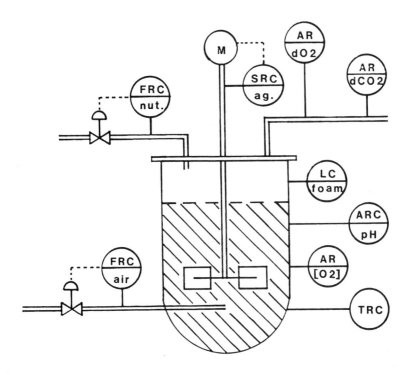

Figure 12. Set-up of the basic laboratory fermentation unit.

(micro)computer system, facilities to expand simply to a non-basic laboratory fermentation unit.

Capacity of the fermentation system for process interfacing is:

(i)   Seven analogue-to-digital converters (ADCs) for input signals from dissolved oxygen tension, off-gas carbon dioxide concentration, off-gas oxygen concentration, air flow, stirrer speed, temperature and pH.

(ii) One digital input (DI) for input signal from level of foaming. When analytical instruments for off-gas analysis are shared, two additional DIs are necessary for indication of the fermentation unit.

(iii) One data interface RS232 for analytical balance of feed tank, when liquid flow is controlled through the decrease of the weight of the feed tank.

(iv) Nine digital-to-analogue converters (DACs) for output signals for control valve air flow, stirrer speed, pump for nutrient flow, and six output signals for an analogue recording instrument.

(v)   Five digital output (DOs) for output signals for pumps of acidic and alkaline titrant and antifoam agent, cooling and heating. Assuming a pause/pulse control of pH and temperature.

The above-mentioned demands for a fermentation unit can easily be met by commercially available fermentation units from well-known suppliers with the sole exception of a reliable and accurate closed-loop control of liquid flow (nutrient).

4.2 Non-basic laboratory fermentation unit

Where as the basic laboratory fermentation unit shows little use of computer technology for the fermentation system, for a 'state of the art' multipurpose laboratory fermentation unit more features are required. One could define a standard laboratory fermentation unit, which reflects a level of control for multipurpose experiments. The absence of on-line monitoring of variables is characteristic of this fermentation unit. Operator intervention is needed only for the off-line monitored variables.

(i) *Off-line monitored variables.* Flow (cumulative) of antifoam agent, flow (cumulative) of acidic and alkaline titrant, (viscosity), nutrient concentrations, active/total biomass, extracellular biochemicals and metabolites.

(ii) *In-line monitored variables.* Bioreactor broth volume, (optical density), concentration of key-nutrients (e.g. carbon, nitrogen), dissolved oxygen tension, off-gas oxygen concentration, off-gas carbon dioxide concentration, air flow, level of foaming, flow nutrient, stirrer speed, temperature, pH. Nine primary control loops can be specified: volume broth bioreactor, air flow, stirrer speed, pH, temperature, level of foaming, dissolved oxygen tension, flow nutrient, concentration of key-nutrient (e.g. precursor). The actual application of these primary control loops also depends on the mode of operation of the experiment (batch, fed-batch, continuous).

Additional demands on the performance of the fermentation system are: generation of time-dependent setpoint profiles (recipes), setpoint feedback control for primary control loops, user-oriented software-based supervisory control, on-line calculation of oxygen uptake rate, data-logging and data-storage, data-processing with user-oriented software. Some of these functions (in particular data-storage and data-processing) might be shared with other fermentation units and installed on an on-line, stand-alone computer system. If so, the fermentation system should include a communication interface, including protocols, for data transmission.

Capacity of the fermentation system for process interfacing depends on choices made for sensors, actuators and control strategies. A typical example would be as follows:

(i)   Eight analogue to digital converters (ADCs) for input signals from optical density, dissolved oxygen tension, off-gas carbon dioxide concentration, off-gas oxygen concentration, air flow, stirrer speed, temperature, and pH.

(ii) Three digital inputs (DIs) for input signal from level of foaming, off-gas concentration oxygen, oxygen concentration for carbon dioxide.

(iii) Three data RS232 interfaces for analytical balances of bioreactor weight and nutrient feed tank (twice). An additional data interface for manual input of off-line monitored variables (keyboard).

(iv) Four digital-to-analogue converters (DACs) for output signals for control valve air flow, stirrer speed, pump for nutrient flow (twice). An additional one might be essential for a continuous experiment (pump for harvest of fermentation broth).

(v)   Five digital outputs (DOs) for output signals for pumps of acidic and alkaline titrant and antifoam agent, cooling and heating.

Based on standard laboratory fermentation unit, more dedicated fermentation units could be designed for a specific experiment. A dedicated fermentation unit serves a special objective, e.g. scale-down experiments with transient environmental conditions

(17−18). However, the level of control will probably not exceed the level of control for the standard fermentation unit.

Future trends will probably follow two directions: more specific in-line sensors, and more profound control of experiment, process and process variables. The first improves the effectiveness of fermentation research; the second improves the efficiency of that research. The speed of further implementation will be determined by the inherent economics of the process.

4.3 Suppliers of fermentation systems

A market survey of computer systems for fermentation control (19) includes two examples for laboratory application: one is a data-logging project, the other a fermentation system shared by two laboratory bioreactors.

We have surveyed the fermentation systems available on the market in 1986. The objective of our evaluation was different from the published survey. Only laboratory fermentation systems available and in operation in 1986 were accepted for the survey. The possibility of awarding a turnkey project based on specifications to an experienced company was excluded from the survey.

The primary basis of the survey was a fermentation system comparable to the one for the basic laboratory fermentation unit (see section 4.1). Many companies could technically meet this specification, except for the facilities for simple expansion to the standard laboratory fermentation unit (see section 4.2). Only two companies could meet this specification for the fermentation system in 1986: Alfa Laval (Alert 50, Alcom), and Rintekno (MFCS). Both systems are described in literature (20−21). The Alfa-Laval fermentation system is currently referred to as CBC 50 and Alcom. The second fermentation system was originally developed for pilot plant bioreactors. An update for laboratory fermentation equipment is currently available from B.Braun (exclusive rights from Rintekno Oy). Both fermentation systems are configurable, i.e. hardware and software are adjustable to user specifications.

However, many companies are developing additional hardware and software that will meet the specifications for the standard fermentation unit. Currently many suppliers offer a fermentation system that fulfils many of the functions required for the standard fermentation unit. The availability of these fermentation systems has unfortunately come too late in comparison with what has been achieved by the fermentation industry (4,22−24) and academic research laboratories. Since a fermentation unit is assembled with parts supplied from several manufacturers and the fermentation unit evolves by adding new features, there is a need for 'standardization' (e.g. remote control of primary controllers by analogue signals and not by 'firmware' data interfaces).

5. ACKNOWLEDGEMENT

We are grateful to H.J. de Groot and E.Nisenfeld for their discussions on control engineering, and all engineers and technicians operating, maintaining and improving the laboratory fermentation units at Gist-brocades Delft, The Netherlands.

Table 4. List of symbols.

Symbol	Standard instrument identification First letter	Succeeding letter	Non-instrument identification
A	Analysis	Alarm	–
C	Conductivity	Controller	–
$[CO_2]$	–	–	Dissolved carbon dioxide concentration
D	Density	Differential	–
dCO_2	–	–	Off-gas carbon dioxide concentration
dO_2	–	–	Off-gas oxygen concentration
E	Voltage	–	–
F	Flow	Fraction (ratio)	–
F_A	–	–	Flow (acidic titrant)
F_B	–	–	Flow (alkaline titrant)
F_G	–	–	Flow (air)
F_N	–	–	Flow (nutrient)
F_S	–	–	Flow (antifoam)
F_W	–	–	Flow (fermentation broth)
H	Hand/manual	–	–
I	Electrical current	Indicator	–
J	Power	–	–
L	Level	–	Level of foaming
M	Humidity	–	–
$[N]$	–	–	Nutrient concentration
$[O_2]$	–	–	Dissolved oxygen concentration
P	Pressure	–	–
$[P]$	–	–	Product concentration
pH	–	–	Hydrogen ion concentratrion
Q	–	Cumulative	–
S	Speed (stirrer)	–	–
T	Temperature	–	–
V	Viscosity	–	–
W	Weight	–	–
$[X]$	–	–	Biomass concentration

6. GLOSSARY

Actuator   An instrument that alters the value of a monitored variable in response to the output signal of a controller or manual intervention of the operator.

Analogue   Adjective referring to continuous variation of a physical quantity (e.g. electrical current).

ADC   Analogue-to-digital converter. A device for the conversion of an analogue electrical signal into its digital equivalent.

BCD   Binary coded decimal: a coded numeric number.

Bit   Binary digit. Basic element of the binary representation of information: 0 or 1. Digital information is based on the combination of (mostly eight) bits.

Byte   Eight bits.

Controller   A device that has a variable output to change the monitored variable in a specified way.

DAC Digital-to-analogue converter. A device for the conversion of a digital value into its analogue electrical equivalent.

Device Piece of electrical, pneumatic, or mechanical equipment (or a combination of these).

DI Digital-input converter. A device for the conversion of a discrete signal into its digital equivalent (on or off).

Digital Adjective referring to coded numeric data.

Discrete Adjective referring to an on- or off- condition.

DO Digital-output converter. A device for switching on or off an actuator.

Instrument A device used to measure and/or control a variable.

Instrumentation The application of instruments.

Interface A device of a computer system that receives information, if required modifies the information or its form or both, and produces a resultant output. Simply, a device for coupling a computer to its environment.

OUR Oxygen uptake rate, i.e. microbial oxygen consumption.

Parameter Adjustable value in a model, control algorithm, or controller considered as a constant.

Relay An electric switch that is actuated by an electrical signal.

Sensor An instrument that first senses the value of the monitored variable.

Sequential control A logic series of action steps performed by an automaton for the operation and control of a (part of a) process (*not* the control of the process variables!). E.g., for fermentation: sterilization routine, cleaning in place procedure, medium preparation, inoculation procedure, calibration procedures.

Standard Generally accepted. 'The nice thing about standards is, that you have so many to choose from; if you do not like any of them, you can just wait for next year's model' (Tannenbaum).

Transducer A device that receives information in the form of a signal, if required, modifies the information or its form or both and produces a resultant output.

Transmitter A device that senses the monitored variable through the medium of the sensor, and that has an output signal varying only as a known function of the variable.

Variable Any property of the process.

7. REFERENCES

1. Solomons,G.L. (1969) *Materials and Methods in Fermentaiton,* Academic Press, London.
2. Armiger,W.B. (ed.) (1979) *Biotech. Bioeng.* Symp No. 9, Wiley, New York.
3. Halme,A. (1983) *Modelling and Control of Biotechnical Processes,* Pergamon Press, Oxford.
4. Carleysmith,S.W. and Fox,R.I. (1984) In A.Mizrahi and A.L.van Wezel (eds.), *Adv. Biotech. Proc.* Alan R.Liss Inc., New York, Vol 3, 1.
5. Johnson,A. (1986) *Modelling and Control of Biotechnological Processes,* Pergamon Press, Oxford.
6. Bushell,M.E. (ed.) (1988) *Progress in Industrial Microbiology,* Vol 25: *Computers in Fermentation Technology,* Elsevier, Amsterdam.
7. Schügerl,K., Lübbert,A. and Scheper,T. (1987) *Chem. Ing. Techn.,* **59**, 701 (in German).
8. Mattiasson,B. (1987) In O.M.Neijssel,R.R. van der Meer and K.Ch.A.M.Luyben (eds), *Proc. 4th Europ. Congr. on Biotechnol.,* Elsevier, Amsterdam, Vol 4, p. 255.
9. Kemblowski,Z. and Kristiansen,B. (1986) *Biotechn. Bioeng.,* **28**, 1474.
10. Picque,D. and Corrieu,G. (1988) *Biotech. Bioeng.,* **31**, 19.
11. Fox,R.I. (1984) In A.Wiseman (ed.), *Topics Enzyme Fermentation Biotechnol.* Ellis Horwood, London, Vol 8, 126.
12. Memmert,K., Uhlendorf,R. and Wandrey,C. (1987) *Chem. Ing. Tech.,* **59**, 501 (in German).

13. Cooney,C.L. and Mou,D.G. (1981) Paper presented at the 3rd Conf on Computer Applications in Fermentation Technology, Soc Chem Ind, London, 217.
14. Brown,D.E. (1982) *J. Chem. Tech. Biotechnol.*, **32**, 34.
15. Working Party 'Bioreactor Performance' of the European Federation of Biotechnology (1984) *Process Variables in Biotechnology*, Dechema, Frankfurt am Main.
16. Working Party on Bioreactor Performance of the European Federation of Biotechnology (1987) *Physical Aspects of Bioreactor Performance*, Dechema, Frankfurt am Main.
17. Oosterhuis,N.M.G. (1984) Thesis, University of Technology, Delft, Holland.
18. Sweere,A.P.J. (1988) Thesis, University of Technology, Delft, Holland.
19. Fox,R.I. (1981) Paper presented at the 3rd Conf on Computer Applications in Fermentation Technology, Soc Chem Ind, London, 91.
20. Wallberg,C. (1987) In *The World Biotech Report 1987* Vol 1. *Europe*: The Proceedings of Biotech 87 held in London May 1987 part 3, On Line, London.
21. Jokela,J., Oinas,R. and Meskanen,A. (1981) Paper presented at the 3rd Conf on Computer Applications in Fermentation Technology, Soc Chem Ind, London 85.
22. De Visser,G.J. (1984) In E.H.Houwink and R.R.van der Meer, *Innovations in Biotechnology*, Elsevier, Amsterdam, 293.
23. Alford,J.S. (1981) Paper presented at the 3rd Conf on Computer Applications in Fermenation Technology, Soc Chem Ind, London, 67.
24. Valentini,L., Andreoni,F., Buvoli,M. and Pennella,P. (1981) Paper presented at the 3rd Conf on Computer Applications in Fermentation Technology, Soc Chem Ind, London, 175.

CHAPTER 8

Sterilization

C.A.KENT, M.K.DAWSON, D.C.HEARLE and D.J.WEALE

1. INTRODUCTION

Whether the purpose is for academic research, process or product development, or some other use, laboratory fermentations will generally be carried out in a controlled environment. This must provide for the growth of the desired organism or organisms and the production of a desired product. Sterilization is usually a vital part of the overall fermentation procedure enabling these objectives to be met. In this chapter, we shall outline the range of sterilization options available and factors affecting the choice of an option, along with the major procedures to be followed in preparing for and carrying out the sterilization. Whereas the scope of the applications envisaged is intended to be as wide as possible, the emphasis of this chapter will be on monoseptic microbiological fermentations, or at least those involving a defined microbial population.

1.1 Why sterilization?

If a fermentation is to be conducted using a defined organism for a desired product or range of products, invasion by contaminating microorganisms will adversely affect it in one or more of a number of ways. Thus, the contaminants will compete with the desired organisms for nutrients, thereby reducing yields; they may even outgrow and displace the desired organism: contaminants will also produce their own products which will have to be separated from the desired products and may even degrade the latter. Therefore, contamination must be avoided. This may be done by:

(i)   Using a pure inoculum.
(ii)   Sterilization of the medium, fermenter, ancillary vessels, and all materials and surfaces coming into contact with these.
(iii)   Maintaining aseptic conditions during the fermentation.

Sterilization and the maintenance of aseptic conditions sometimes leads to the loss or degradation of sensitive medium constituents. The design of the most suitable sterilization procedure for a given application will therefore usually be a compromise based upon the two basic aims of sterilization (1):

(i)   To attain the desired probability of sterility (and to maintain this probability of non-contamination throughout the fermentation).
(ii)   To avoid or minimize any loss of medium nutritive quality caused by sterilization.

In deciding upon the nature of this compromise, we are faced with a number of questions which we shall consider in this chapter. One of the earliest questions we shall

173

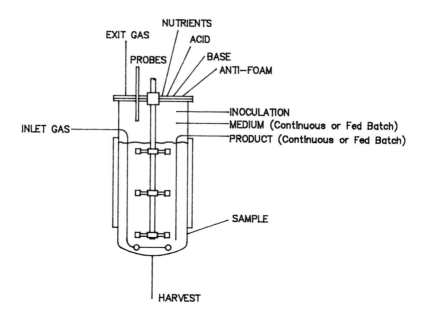

Figure 1. Generalized fermenter flowsheet.

need to answer is, 'How likely is my fermentation to be contaminated, and how important would it be?', that is, we must consider what our particular 'sterilization' criterion should be.

1.2 Criteria for sterilization

Avoiding contamination of a fermentation system is usually important, even vital. However, in general, measures are used that are no more severe than necessary, to avoid damage to medium and reduce cost. Therefore, the 'sterilization' procedure adopted and the extent of removal of contaminants need to be related to the nature of the fermentation and its purpose.

Thus small-scale work on waste water treatment, say using an 'activated sludge' oxidation, will use the organisms naturally present. Equipment for such work would probably not be sterilized except, perhaps, for vessels and lines containing nutrients. Some single-organism fermentations are described as 'protected': either the medium will support the growth of a limited range of organisms only (e.g. a hydrocarbon-based medium) or the organism produces an environment hostile to other organisms (e.g. ethanol production by yeast). In such cases, a less extreme regime might be employed, perhaps involving cleaning vessels and lines and use of disinfectants, and boiling or pasteurization of the medium, which will kill most non spore-forming organisms but not all spores (2).

However, most laboratory fermentation work is likely to involve 'non-protected' pure-culture fermentations. For these, essentially all microbial contaminants must be excluded. This ideal of total removal is, however, far from easy to attain. Therefore, the criterion of 'probability of contamination' is frequently used as a more practical aim. A probability

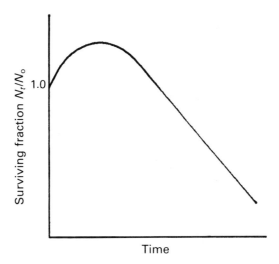

Figure 2. Effect of heat on spore survival.

commonly acceptable in industrial fermentations is 10^{-3}: that is, the probability that 1 in 1000 batches will be contaminated after sterilization. More or less stringent requirements may be imposed depending upon the consequences of failure. An example from another area is that of canning of food, where a sterilization design criterion of a *Clostridium botulinum* spore survival probability of less than 10^{-12} is typical (3). A comparable fermentation application might be where dangerous pathogens or certain products of recombinant DNA technology are involved.

The choice of the sterilization method and regime to be used must therefore be made in the light of the appropriate criterion of sterilization. This needs to be decided upon at an early stage.

1.3 Areas for sterilization

Having decided upon the stringency required for our sterilization operations, we now go on to consider what it is, in our fermentation system, that we have to sterilize. *Figure 1* shows a generalized 'flowsheet' centred on a fermenter. It includes those material flows, lines, and equipment likely to require sterilization, and is intended to be of general application, covering batch, fed-batch and continuous fermentations, with or without pH and chemical foam control. The vessel type featured is the aerated stirred-tank fermenter, which is the 'standard' type of laboratory fermenter. However, the same principles will apply to other fermenter types.

Sterilization is likely to be carried out in four types of area:

(i)   The fermenter itself and associated vessels (such as reservoirs and receiving vessels).

(ii) Liquids, either in the fermenter or in other vessels or lines: the medium, acid and alkali for pH control, antifoam, and any additional nutrients or reagents.

(iii) Gases: notably air for aerobic fermentations.

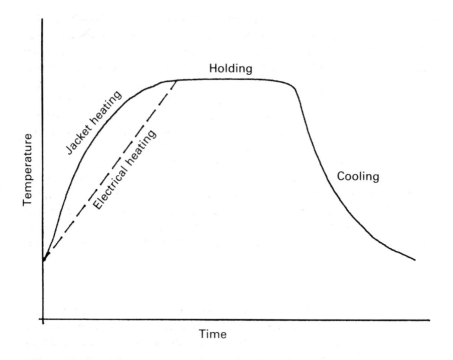

Figure 3. Batch sterilization temperature–time profiles.

(iv)   Lines, valves, and peripherals: including probes, shaft seals, pumps, and line couplings/connections.

   Each area will be dealt with separately. However, some form of sterilization is likely to be required at all stages of the fermentation process: i.e. preparing for fermentation, during fermentation, and after fermentation. This must be planned for, and suitable sterilization procedures adopted in each case.

1.4 Methods of sterilization

Contaminating organisms may be removed from a system in a number of ways, each differing in its effects, ease of use, and effectiveness. At present, the options which are, or could be, available to us on the laboratory scale include:

1.4.1 *Moist heat*

Sterilization using moist heat usually involves steam at temperatures above 100°C, most commonly, at 121°C. It is usually carried out either in a pressurized autoclave or via steaming *in situ*. Thermal death of microorganisms by this means is thought to be due to denaturation of enzymes and degradation of nucleic acids in the cell. As such, it is generally a first-order rate process (1). However, if heating is insufficient, many types of spore may, in fact, be activated and start multiplying. This effect is illustrated in *Figure 2*, where: N_t is the spore population after heating time, t, and N_0 is the initial spore population (see also refs 4−6).

Instances of activation conditions for bacterial spores are given by Russell (6). For example, certain strains of *Bacillus stearothermophilus* are activated after 7−10 min at 110 to 115°C. Therefore, it is essential that the sterilization regime should provide every volume of fluid and every surface with a sufficiently long exposure time at a *lethal* temperature. Time for heating every part of the system up to this temperature must be allowed for in the design of any sterilization procedure. For example, if a fermenter is to be autoclaved, the temperature in the centre of the medium it contains will be slowest to rise, and will lag behind the temperature in the autoclave itself. Therefore, time needs to be allowed for the temperature in the medium to approach that of the autoclave. This is usually done by extending the 'holding' time set. When calculating sterilization conditions, the heat-resistant spores of *B.stearothermophilus* are usually taken as the target, or 'worst-case', contaminant. Various methods exist for the design of heat sterilization conditions—for example, the 'Del factor' approach (see refs 1, 4, and 5) but these are outside the scope of this chapter. This option is characterized by a sterilization cycle comprising three phases: heating, holding, and cooling (see *Figure 3*).

1.4.2 *Dry heat*

Sterilization using dry heat is thought to be caused by reaction between two compounds, such as the oxidation of cellular proteins and nucleic acids. It is usually effected by means of hot air in an oven, although use may be made of inert gas, vacuum, infrared radiation, superheated steam, or even hot vegetable oil (6). The basis of this means of sterilization is to maintain the object to be sterilized at the desired temperature for the requisite period of time, as with moist heat. However, dry heat is less effective in destroying contaminating organisms, so higher temperatures (often 180°C or higher) and longer times than for moist heat sterilization must be employed. Furthermore, the different means of dry-heating have different levels of effectiveness: microorganisms are more resistant to hot gases of low water content than to superheated steam.

1.4.3 *Filtration*

Both liquids and gases may be sterilized by passage through various types of microbial filter, which will retain organisms present and other solids of similar size or larger. Gases may be sterilized using either membrane filters or depth filters, both of which are discussed in more detail in Chapter 1. Filter sterilization of liquids is accomplished using membrane filters.

1.4.4 *Chemical means*

Destruction of contaminating organisms to various extents may be achieved by means of disinfectants or sterilizing agents. Such agents may be liquid-phase or gas-phase. Liquid-phase agents include chlorine compounds, ethanol, hydrogen peroxide, formaldehyde solutions, and sodium metabisulphite. Gas-phase agents include ethylene oxide and formaldehyde vapour. Care must be taken, however, in matching the agent to the system and the degree of contaminant removal required. Also, certain agents must themselves be removed subsequent to the sterilization so that the desired fermentation may go ahead, for example, by flushing with pre-sterilized fluid.

1.4.5 *Radiation*

Ionizing and ultraviolet radiation have found increasing sterilization applications, primarily for instruments, filters, etc.

Currently, moist heat is the most common sterilization option used for vessels, lines, and liquids, with filtration the common choice for gases and very heat-sensitive medium constituents.

2. STERILIZATION OF FERMENTERS AND VESSELS

In this section, we shall consider the fermenter and its associated vessels together. Whether they are to be sterilized full of liquid or empty, the same factors must be considered for all of these vessels. Before discussing the practicalities of sterilization, we shall look at how to choose a suitable method.

2.1 Choosing a sterilization option

As we have already seen, there are a number of methods of sterilization, with moist heat being the most common means for vessels. Although we shall concentrate on this, there are various ways of introducing moist heat, and occasionally other options may be chosen instead. Therefore, we must consider those factors that are important in determining the most suitable means of sterilization. These are:

(i)   The nature of the fermentation and the organism(s) used.
(ii)   The level of containment required: what is the Hazard Group category of the organism?
(iii)   The nature of the medium and the contents of addition vessels.
(iv)   The size and type of the fermenter/vessel; its materials of construction.
(v)   The services available.
(vi)   Is the vessel to be sterilized empty or containing medium/other solutions?

Factors (i) to (iii) have already been considered, or will be considered in later sections. In this section, therefore, we shall only look at the last three points.

2.1.1 *Fermenter/vessel size and type*

By far the most common type of laboratory fermenter is the stirred, aerated tank, so we shall concentrate on this. However, some workers may wish to use other designs, such as air-lift and bubble-column fermenters, so brief consideration will be given to these. Nevertheless, similar principles apply to all types, although the constraints may be different.

(i)   The vessel size and the materials used for its construction affect the sterilization method used. Thus, the two usual ways of introducing moist heat (autoclaving and *in-situ* steaming) will be more or less suitable, depending upon vessel design and materials. Vessels larger than approximately $15-20$ litres working volume are not generally portable because of their size and weight. These must therefore be capable of undergoing *in-situ* sterilization. If steam-sterilization is to be used, as is normal, the materials of construction must be able to withstand the stresses involved. Moist heat sterilization is generally carried out at 121°C and 15 p.s.i.g.

(103.4 kPa) steam pressure, so glass vessels above this size will not normally be strong enough. Therefore, these small pilot-scale fermenters will almost always be constructed out of type 316 stainless steel, or equivalent, as is the case with most stirred-tank vessels. Where a larger fermenter must be made of glass, for example for flow visualization in air-lift or bubble-column fermenters, a form of sterilization may be accomplished by careful steaming out for several hours at atmospheric pressure. This may be preceded by chemical cleaning and sterilization. *In-situ* steaming of pilot-scale stirred-tank fermenters is most commonly done by passing saturated steam through an external jacket and directly into the vessel. Some smaller vessels in this size range may be associated with their own steam-raising equipment. Others may be provided with an electrical heater, such as a band heater, between the vessel and the jacket.

(ii)   Stirred tank fermenters smaller than approximately 15−20 litres are most commonly of composite construction (Chapter 2), using a glass body and stainless steel top/bottom plate(s). A few laboratory fermenters are constructed entirely from stainless steel. The preferred method of sterilization of composite glass/stainless steel vessels is autoclaving, so it is important to ensure that a suitable autoclave is available (Chapter 1). There are composite fermenters on the market that are provided with *in-situ* steaming facilities. If such a vessel is to be used, it is essential that it be fitted with a strong steel guard, preferably of solid sheet rather than mesh. This should be close-fitting and adequately designed to protect operators against flying glass splinters produced if breakage occurs during pressurization. Manufacturers' systems should be assessed carefully, as a failure in this area could cause serious injury. *In-situ* steaming may be effected by use of mains steam or steam raised by electrical heating. Small stainless steel fermenters will normally be sterilized *in situ*.

(iii) The same factors must be considered in the cases of addition vessels for smaller fermenters, which themselves are likely to be either of composite or solely glass construction, and other types of fermenter. Glass vessels to be heat-sterilized must always be autoclaved: the larger the vessel diameter, the less stress it will take. Glass air-lift and bubble-column fermenters are likely to be long in relation to diameter, so may be difficult to fit into an autoclave. Again, atmospheric pressure steaming for several hours may be the only viable alternative. In this case, as with larger fermenters, the design and steaming procedures must be such as to positively prevent pressure build-up. An in-place guard is worth considering even here.

2.1.2 *Services available*

The sterilization option chosen for a fermentation will depend upon the type, quantity, and quality of services available. Therefore, at an early stage, the services required for a desired sterilization method, and any associated equipment (such as autoclaves), should be compared with those services that can be made available (Chapter 1).

2.1.3 *'Full' or empty sterilization?*

The common practice with stirred-tank fermenters is to carry out the sterilization with the vessel containing medium. A head-space of 30% or more of the total vessel volume is normally needed, to prevent 'boiling over' into the exit air filter. Sterilization in this way is less complicated and less risky than where the vessel is sterilized empty and separately from the medium, and the medium added to the fermenter after sterilization. However, there are instances where it is desirable or necesary to sterilize the fermenter empty. For example, this may be the case with air-lift and bubble-column fermenters, whether sterilization be by atmospheric pressure steaming-out or autoclaving. Atmospheric pressure steaming-out would not be adequate for medium sterilization, so separate autoclaving of the medium is called for. Smaller fermenters of this type, if autoclaved, are likely to be placed at an angle in the autoclave, to enable them to fit in. Therefore, empty sterilization will be necessary in these cases, unless the fermenter can be arranged vertically.

It is sometimes considered desirable to sterilize a stirred-tank fermenter empty prior to batching-up with medium and sterilizing the vessel 'full'. This is particularly so for *in-situ* sterilized fermenters, where the first sterilization, of the empty vessel, serves as a final cleaning stage as well as providing extra sterilization time. Alternatively, empty-vessel sterilization allows the option of different means of sterilization to be used for the fermenter and the medium, for example, filter-sterilization for a heat-sensitive medium.

2.1.4 *Heat sterilization: operating factors*

The rate of heat destruction of microorganisms increases with temperature more rapidly than that for sensitive medium constituents. Therefore, whichever method of heat sterilization is used, it is generally desirable to heat up the fermenter/vessel and any contents as rapidly as possible to the holding temperature, and to cool down as rapidly as possible after the holding phase. It is worth bearing this in mind even when the choice has been made whether to autoclave the vessel or to steam it *in situ*, for there are various options available for autoclaving, and different means of steaming.

(i)   *Autoclaving.* The means of heating the autoclave will often affect the time taken for it to reach the holding temperature, so it is advisable to obtain this information from the supplier if choosing between different autoclaves. As a general rule, for a given means of heating, the larger the autoclave the longer the heating and cooling times. When sterilizing, the holding time should begin only when the sterilization temperature has been reached at the coolest part of the vessel (usually the middle, for a vessel containing medium). In the authors' laboratories, vessels of up to 1 litre are held at 121°C for 15 minutes. Larger vessels are held for 30 minutes, occasionally longer. It must be remembered that temperatures inside the vessel will lag behind that in the autoclave, both during heating and cooling. This will be especially so for glass vessels and those made out of autoclavable plastics, such as polycarbonate, both of which have poor heat transfer properties. Furthermore, heat transfer to and from liquids in vessels will also be slow, since agitation is not normally feasible. The larger the diameter of the vessel, the larger the temperature lag will be. Therefore, the difference between the temperature in the middle of the vessel and that indicated in the autoclave can be considerable. Measurements made in the authors' department (K.Kaur, R.Santos, and R.I.Done, personal communication) illustrate these points. In the same autoclave, 10 litres of medium in a glass aspirator, 10 litres of medium in a polypropylene container, and 5 litres of medium in a glass aspirator, took the following times to heat up to 121°C in the centre of each vessel respectively: 80 min, 160 min (estimated), and 50 min. The empty polypropylene container took 20 min to heat up. Sometimes, volume losses occur, due to evaporation. These are usually small, but need to be considered and checked.

(ii) *In-situ steaming.* Fermenters sterilized in this way are generally heated either by passing steam through the jacket or via electrical heaters. The temperature – time profiles for the heating phase for each of these two means of heating are slightly different; steam through the jacket produces an 'exponential' approach to the holding temperature, whereas electrical heating exhibits a linear approach (3). Typical profile shapes are shown in *Figure 3*. In addition to this, steam is also introduced into the sparger and as many other lines as possible, when the vessel temperature has attained 95 to 100°C. This is for three reasons: to sterilize the lines themselves, to provide positive flow in order to expel all air from the vessel, and to add extra heat to speed up the cycle. Occasionally, all the heating is done by direct steam injection into the vessel. There is likely to be a volume change of the medium during sterilization, due to condensation of some steam during injection and evaporation losses during cooling, as air is usually passed through the vessel to prevent vacuum formation and dry the air filters. The extent of this volume change, and whether it is positive or negative, depends upon the balance between these effects. The lower the temperature at which the steam sparging starts, the greater the volume gain due to condensation, especially below 100°C. A 5 – 10% increase in volume of the medium is not unusual. If the vessel has been sterilized empty, this volume of condensate is likely to be left in the fermenter after sterilization and cooling. Such a volume change must be taken into account when batching-up the medium, to prevent dilution or concentration after sterilization: the medium constituents are made up in a correspondingly smaller or larger volume before sterilization, depending upon whether the nett volume change is positive or negative. Fermenters sterilized *in situ* containing medium are usually agitated, to improve heat transfer and minimize temperature gradients, which will be far smaller than would be the case for autoclaving. Nevertheless, it is often the case that the larger the vessel the slower the rate of heating and cooling. Good design, can however, improve this.

2.2 Preparation and sterilization

Having chosen the means of sterilization, we must then prepare the fermenter and associated vessels and carry out the sterilization. These operations will be discussed here with the emphasis on the use of moist heat, the most common means of sterilizing vessels. However, a brief consideration will also be given to chemical methods.

2.2.1 *Heat sterilization: general points*

When preparing for sterilization, one must consider the whole fermentation system; fermenter, addition vessels, peripheral units, and lines. Preparation should be conducted in a methodical manner, and double-checked because any mistakes will result in either loss of the fermentation or the need for a repeat sterilization.

Initially, the vessel and peripherals should be carefully cleaned using a suitable

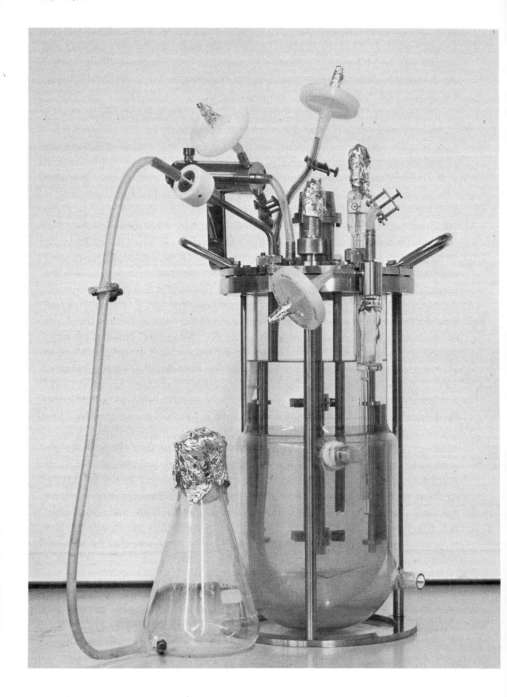

Figure 4. A fermenter prepared for autoclaving.

detergent followed by thorough rinsing with water. In some cases, an empty-vessel sterilization may be considered to be desirable as a final cleaning stage, especially for *in-situ*-sterilized equipment (see section 2.1.3). Following this, if the fermenter is to be sterilized 'full', medium should now be added. Also, acid, base, antifoam and other addition vessels should be filled and attached to the vessel, if these are to be sterilized together. The inoculum vessel should also be connected, if desired. Otherwise, the lines attached to these vessels, and any addition ports on the fermenter, should be clamped and the ends suitably plugged and capped. It is also important to remember not to overfill addition vessels. After filling, all vessel ports must be secured and closed and headplates tightened, to avoid leakage. Where used, new septa should be fitted. These should also be covered by secure caps on vessels sterilized *in situ*, to prevent any escape of steam in the event of septum failure.

During heating, the exit air line should remain open, to allow air to be expelled from the vessel and to allow steam to contact the exit filter. It is important to expel all air from the system to ensure that all areas are exposed to moist heat, for maximum sterilization effect. Furthermore, if the sterilization cycle is controlled by temperature, areas where there are air pockets will be subject to over-pressure. The inlet air filter must, where possible, be isolated from the inside of the vessel, if it is to be autoclaved, to prevent draw-back of medium through the sparger into the filter. The inlet to that filter must be plugged and capped, to prevent wetting, but not itself be clamped. *In-situ* sterilization of pilot-scale fermenters normally provides for steaming-through of the inlet filter independent of the vessel, before steaming into the fermenter. In all cases, it is important to avoid wetting filters during the heating period, as this will not only compromise filter efficiency but may allow vacuum formation in the vessel during cooling. For this reason, the exit from the exit filter must also be plugged and capped if the fermenter is to be autoclaved.

In summary, whether sterilization is to be by autoclaving or *in situ*, all lines to and from the vessel should be closed; in the case of lines in systems to be autoclaved, preferably in two places per line, one near the fermenter, the other near the addition vessel. The exception is any line used for the expulsion of air, such as the exit air line, which should remain open. Addition vessels must be fitted with air vent filters, which should be plugged and capped if they are to be autoclaved. Electrical connections on probes should also be covered with (non-absorbent) cotton wool and capped, as should pockets, prior to autoclaving. Galvanic dissolved oxygen probes may need to be shorted, and antifoam probes isolated before sterilization begins. A photograph of a small fermenter prepared for autoclaving, showing the main precautions to be taken, is given in *Figure 4*. During the cooling phase of the sterilization cycle, steam will condense inside the vessel. This will cause a vacuum to form unless measures are taken to prevent it. Vacuum formation is undesirable, for three reasons: it may damage the vessel, it may cause air to be sucked into the vessel through any ineffective seals or a perforated/collapsed air filter leading to contamination, or it may cause the hot medium to boil over. Therefore it is common practice with *in-situ*-sterilized fermenters to pass air into the vessel through the inlet filter as the temperature approaches just over 100°C. This maintains a positive pressure, if sufficient air is used, and helps to dry out the inlet and exit air filters, which may have collected some steam condensate on cooling.

There is less control over vessel pressure during cooling in an autoclave. However, the rate of cooling should not be so rapid as to cause boiling over of vessels inside still under pressure. Therefore, sudden reductions in pressure should be avoided. On the other hand, the autoclave should not allow vacuum formation to occur. Many autoclaves are fitted with a vacuum-release valve which allows air into the autoclave if a vacuum forms. Some will allow pressure control to allow a small positive pressure to be maintained, but will also be fitted with a safety lock to prevent opening above a set temperature, often 80°C. It is often advisable to remove the fermenter at such a temperature and place it in its rig. The temperature control system is then connected up to give a rapid cooling rate, and a small air flow passed through the vessel via the inlet air filter, to maintain a slight positive pressure and dry out the filters, as with *in-situ* sterilization.

2.2.2 *In-situ steaming of pilot-scale vessels containing medium*

A summary of the procedures followed by the authors in sterilizing a pilot scale 150 litre fermenter with a full charge of medium is given below as an example:

(i) *Services required.*

(a)   *Steam*: Filtered upstream of vessel, preferably regulated to a pressure of 35 to 40 p.s.i.g. (241 to 276 kPa gauge).

(b)   *Cooling water*: Regulated to 25 p.s.i.g. In hard-water areas some water-softening is advisable to reduce scaling of surfaces.

(c)   *Air*: Minimum pressure of 2 bar (29 p.s.i.g.) gauge. Pre-filtered to remove solids, water, and oil droplets which could damage the mass flowmeters.

(d)   *Electricity*; Three-phase, for motor and control panel.

(ii) *Pre-sterilization checks.*

(a)   Open condensate drain valves, and valves on exit air line. Close all other valves.

(b)   Check that the headplate is secure, ports, and fittings are sealed, and the foam probe is at the required level, or lower. Fit new septa and secure protective caps over these, and blank end on the inoculation port.

(c)   Pressurize the pH probe to 1.5−2.0 bar.

(d)   Drain condensate from steam lines. Open inoculum valve.

(e)   Isolate all non-sterilizible equipment.

(iii) *Stages of the sterilization cycle.*

(a)   *Ambient to 50°C:* Drain the fermenter jacket of water and pass in the steam. Begin heating to 121°C, with vessel agitation to aid heat transfer. Expel air from the vessel via the exit air line and filter and the inoculum port.

(b)   *From the start*: Freely steam through the inlet air filter, harvest and sample valves and pH probe housing. Periodically drain any condensate from the inlet air filter.

(c)   *50 to 100°C*: Close the jacket drain and heat the vessel via steam through the jacket.

(d)   *100 to 110°C*: Pressure is developing in the vessel. Pass steam through the sparger into the vessel and also directly into the headspace, to aid air expulsion and to sterilize the sparger and headspace lines. (As an alternative, this may be done from 95°C. If any other inlet lines are fitted, these should be steamed out also at this stage.) Steam leaves the vessel via the exit air line and filter and the inoculum port through steam traps. At this stage, monitor the vessel temperature and pressure carefully; over-pressure, especially below 100°C, may indicate the presence of residual air or volatiles from the medium.

(e) *110 to 121°C*: Reduce steaming through the sparger to a bleed, and cut off the steam supply to the headspace, to prevent overshooting the holding temperature of 121°C. Periodically drain any condensate from the inlet and exit air filters and exit air line catchpot.

(f) *121°C*: Hold the vessel temperature at 121°C for 30 min (or as determined otherwise).

(g) *121 to 110°C*: Shut off the steam supply to the jacket, sparger, headspace, and inlet air filter. Close the inoculum valve and pass steam through its blanked-off end. Vent steam from the jacket and allow cooling water to pass once-through the jacket and out to drain, for rapid cooling.

(h) *110°C to setpoint*: Allow air to enter the vessel headspace, to dry out air filters and prevent vacuum formation. Control the vessel pressure at a slowly falling level, approximating to just over the saturation vapour pressure of steam at the measured temperature, to prevent boiling; roughly, 10 p.s.i.g. (69 kPa gauge) at 110 to 112°C, and 5 p.s.i.g. (34.5 kPa) at 100°C. This is effected by throttling the exit air line valve. Below 100°C, the pressure is maintained at no less than 3 p.s.i.g. (20.7 kPa). Periodically drain any condensate from the air filters and catchpot. Below approximately 70 to 80°C, pass cooling water through the exit air line condenser.

(i) At setpoint temperature: Switch on the jacket circulation pump and maintain a feed-and-bleed water supply. Maintain a vessel head pressure of 2 to 5 p.s.i.g. (13.8 to 34.5 kPa gauge).

Figure 5 shows a representative temperature−time sterilization profile obtained using this procedure and a steam mains pressure of 40 p.s.i.g. (276 kPa gauge). The effect of reducing the steam mains pressure to 25 p.s.i.g. (172 kPa gauge) is also illustrated.

2.2.3 *In-situ steaming of empty pilot-scale vessels*

Empty-vessel steaming is generally used as a cleaning and pre-sterilization step, before batching and full sterilization. The object is to steam through all parts of the fermenter, pipework, valves, and fittings to bring them to a clean, hygienic state. The procedure is similar to that for sterilizing a vessel containing medium, with the following exceptions:

(i) No steam is passed into the jacket. The jacket drain valve is open until the end of the 'hold'.

(ii) Steaming through the sparger and headspace line, via the inlet air filter, is carried out from the start of the cycle.

(iii) Steaming is also carried out, from the start, through the inoculum valve into the vessel.

(iv) If possible, the drain line is blanked off and steaming is carried out, from the start, through the drain valve and into the vessel.

(v) The vessel is held at 121°C for only 10 to 15 min. During this time, steam to

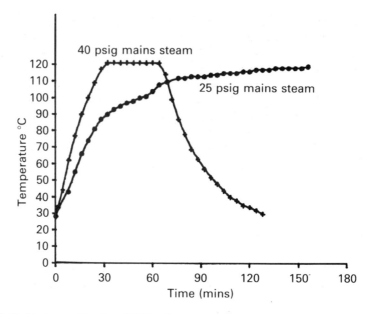

Figure 5. Sterilization profiles for a 150 litre fermenter.

the sparger, headspace, and drain and inoculum valves is throttled back to avoid overshooting the holding temperature.

(vi) After the 'hold' period, the steam supply is shut off to the sparger, headspace, and inlet air filter, and air is passed through the inlet air filter to condensate drain, slowly, to dry the filter.

(vii) Shortly afterwards, steam is shut off from the inoculum, drain, and sample valves, and the inoculum and sample valves are themselves closed, along with the jacket drain valve. Cooling of the vessel is started, with once-through flow of cooling water through the jacket, and the procedure follows stages (h) and (i) of section 2.2.2 (iii).

After this cycle, the vessel will contain perhaps 10–20 litres of residual steam condensate (in a 150 litre vessel). This may either be drained off before batching, or may be used as 'make-up' water for the medium.

2.2.4 *Chemical sterilization*

The use of chemical means of sterilizing fermenters and other vessels is far less common than the use of moist heat, and is generally much less effective. It may be necessary to employ a chemical method if the vessel to be sterilized contains material that will not stand up to the conditions of heat sterilization, but such cases are likely to be rare. However, if the vessel configuration makes it awkward to fit in an autoclave or unsuitable for steaming out *in situ*, and the fermentation does not require a high level of contaminant removal, such as with 'protected' fermentations, then chemical sterilization may well be the most appropriate route. Furthermore, it may be considered desirable to chemically 'sterilize' and clean a vessel before heat-sterilizing it, as an added precaution.

Sterilization/disinfection regimes vary in detail, but always involve a period of contact of the vessel with the sterilant, at ambient temperature or above. Following this, removal of the sterilant is usually required, via flushing or rinsing, unless the sterilant itself decomposes during the contact period. Such rinsing may need to be done using a pre-sterilized rinsing fluid, to maintain 'sterility'. Sometimes it may be desirable to start the process off with a detergent clean. A typical 'cleaning in-place' (CIP) cycle, such as that used in brewing and food processing operations, is given below, as a rough guideline to the procedure which might be used with a liquid-phase sterilant or disinfectant. For the sterilization of laboratory-scale vessels, some modifications may be necessary, such as using agitation rather than circulation:

(i)   Water pre-rinse; ambient temperature for 3 to 5 min.
(ii)   Detergent circulation; ambient to 90°C for 5 to 45 min.
(iii)   Water rinse, to remove detergent; ambient temperature for 3 to 5 min.
(iv)   Sterilant circulation; usually ambient temperature for 1 to 15 min.
(v)   Final water rinse, to remove sterilant; ambient temperature for 3 to 5 min.

Extensive information on the use and effects of liquid-phase and gas-phase sterilants is given by Russell (6), who also discusses other sterilization methods.

3. STERILIZATION OF LIQUID MEDIA

The effectiveness of a sterilization regime for liquid media depends not only on that regime itself, but on the method of medium preparation, and the interaction of both with the medium components. We shall consider each of these areas in this section.

3.1 Medium preparation

The purpose of medium preparation is to ensure the consistency of the operating procedure and thus to maintain the best conditions for the process (see ref. 7).

Medium preparation includes the storage and handling of medium components. Media should be stored under cool, dry and clean conditions. The efficacy of any sterilization cycle is dependent on the assumption of an initial level of contamination in the medium. If storage conditions are such that growth of organisms is favourable the medium may not be sufficiently treated during the designed sterilization cycle, and contamination may result. In addition, the nutritional value of medium components will be affected if microbial growth is allowed to occur during storage.

When preparing a medium it is important to be clean and tidy. If the medium is prepared in an area close to the site of the fermentation it is important to realize that spillages of nutrient components will allow the growth of organisms that are present in the environment. As a result of this, areas may be heavily contaminated with spores that may compromise the designed sterilization regime. It is therefore of paramount importance that medium preparation areas are kept clean and spillages are immediately dealt with. Chemical disinfectants may be used as an aid to control contaminants, and medium preparation areas should be regularly swabbed down with a suitable disinfectant (e.g. Hibitane from ICI).

It has been stressed that sterilization cycles assume an initial level of contamination, and because of this it is important to reduce possible sources of contamination. Vessels used for fermentation, including feed vessels and any lines open to the vessel, should be regularly cleaned and any caked-on deposits of medium (which may provide a haven for organisms during sterilization) should be removed. A strict cleaning schedule is important if fermentations are not to be lost as a result of contamination. As a general rule, medium should be sterilized as soon as possible after it has been made up.

In preparing a medium, it is important to have a standard method of operation in which the order of addition of components is strictly adhered to. The reason for this is that in making up a medium which may involve inorganic salts, oils, nitrogenous compounds, and carbohydrates, complex chemical and physical reactions may occur, such as precipitation reactions, adsorption reactions, the evolution of gases (for example, mixing ammonium sulphate with chalk will cause the evolution of ammonia) and the creation of undesirable pH profiles. Knowledge of the chemistry of the medium is therefore important. Many medium components are powders and these should always be well dispersed before sterilization or else lumps will form through which heat transfer will be inefficient. If oily components are added to a medium before a powder is fully suspended or dissolved, the powder may preferentially dissolve into the oil phase and cause the formation of gel-like lumps during subsequent heat treatment. This not only poses a sterility risk, but also depletes the medium of nutrients.

If a medium component is to be used in which there is a known high quantity of contaminant naturally present (as in frequently the case with molasses), this should be batched into the medium last. Doubling times of bacteria are of the order of $1-2$ h or less, and if a significant time lapse occurs between medium preparation and the initiation of the sterilization cycle, significant growth of the contaminant may occur which may render the cycle ineffective, or at the very least will reduce the nutritional value of the medium, and may introduce toxins.

During heat sterilization pilot-scale 'sterilize-*in-situ*' vessels are generally agitated. However, most laboratory equipment is sterilized by autoclaving and agitation is therefore not possible. Because sterilization is essentially a cooking process, if the medium has not been properly mixed, stratification will occur where different reactions have taken place at different levels in the vessel during sterilization. This may be avoided by intense mixing before the vessel is autoclaved.

There may be advantages in sterilizing medium components separately to avoid degradation reactions and loss of heat-labile growth factors. However, this does increase the number of aseptic transfers required, and may increase the number of vessels and lines to the main vessel, making equipment more cumbersome to handle.

The aim of medium preparation, as was stated at the start of this section, is to have a consistent operating procedure so that each batch is as similar as possible. Clearly there may be variability in the medium components themselves, particularly if the component in question is of a complex nature (for instance molasses). Accordingly it is worth knowing the supplier and the consistency of the composition of the medium used.

3.2 Types of medium component

3.2.1 Carbon sources

The choice of carbon source at laboratory scale depends largely upon physiological factors. Process economics tend to have an increasing influence on the selection for large-scale fermentation processes. Often the carbon source will be a carbohydrate, and there are specific problems associated with sterilization of carbohydrates. During sterilization, reducing sugars (for example, glucose) will react with other compounds in the medium, giving rise to toxic products that will inhibit growth. A typical reaction of this nature is the Maillard reaction between sugar and nitrogenous compounds (for example amino acids). To avoid these reactions it is good practice to sterilize the carbohydrate separately from the other medium components. When carbohydrates are sterilized in solution, caramelization may become a problem. The characteristic browning represents the formation of toxic products. Caramelization may be avoided by lowering the pH to 4 for sterilization of the carbohydrate.

The choice of carbohydrate is generally restricted to the following four:

(i)   *Glucose*. This may be supplied as anhydrous glucose, of various degrees of purity, and glucose monohydrate produced by starch hydrolysis, which is about 95% pure. It is also available in a less refined form as glucose chips, a solid of about 75% glucose. There are also glucose syrups, which are fairly pure but are difficult to handle owing to high viscosity, and which, being about 60% water, are bulkier to store than the solid forms. The high osmotic pressure provides protection from gross contamination during storage, and further protection can be achieved by reducing the pH to $4-5$.

(ii)   *Starch*. This is usually prepared from maize, wheat, or potatoes. It may be bought as soluble starch (starch partially hydrolysed with acid or α-amylase) so that it will more readily form a colloidal suspension. Incorrect batching of starch will lead to the formation of large solid aggregates in the medium, and these will be insufficiently sterilized due to poor heat penetration. This may be avoided by mixing the starch to a slurry before addition or, alternatively, heat treatment of the soluble starch with bacterial α-amylase to further break down the molecules.

(iii)   *Sucrose*. This is available in three forms: white sugar, brown sugar, and molasses. The first two are solid crystalline products of different degrees of refinement. Molasses is a very crude preparation but is probably the cheapest carbohydrate source available. Typically molasses has about 50% sucrose and a variable nitrogen concentration from $0.5-2\%$ depending on the source. The nitrogen component of molasses has two consequences for sterilization. Firstly, the sugar cannot be separately sterilized from the nitrogen component and therefore browning reactions cannot be avoided. Secondly, the nitrogen component in the form of peptides makes any molasses-based medium susceptible to foaming. To reduce foaming it is worth batching antifoam to the medium prior to the sterilization.

(iv)   *Lactose*. This is the only common sugar from animal sources. It is metabolized slowly by many microbes and for this reason was used for fermentations where secondary metabolites were the product, notably penicillin fermentation. The same effect can be achieved by using a glucose feed, and consequently lactose is not commonly used today.

Carbohydrates are not the only possible carbon sources. Others include oils (for example, maize oil), organic acids (for example, citric acid), hydrocarbons, and alcohols. When carbon sources are used that are immiscible with water (and this also applies to antifoams), adequate dispersion of the oil in the aqueous phase is essential to ensure that oil droplets are small and therefore are sufficiently sterilized. Ideally, steam should be sparged through the oil with agitation in order to sterilize it. As this is not always possible, it may be worth sterilizing an oil component by a method other than heat, or to use a much more intense sterilization cycle based on dry heat destruction of spores.

3.2.2 *Nitrogen sources*

About 10% of dry cell weight is nitrogen. Many organisms will use inorganic nitrogen, and the choice will have a profound effect on the pH profile of the fermentation. For instance, ammonium sulphate will lead to the formation of sulphuric acid on the assimilation of the ammonium ion, and consequently the pH of the broth will fall during the course of the fermentation. In general most inorganic sources will not pose a sterilization problem.

It is not uncommon to find that the highest yields are obtained with complex nitrogen sources which contain peptides, amino acids, and many other compounds that may act as inducers for the production of the material of interest. Typically they are supplied as fine powders or meals. If a coarse meal is to be added it should be milled to reduce the particle size before heat sterilization is used. Many such sources can be quite crude and contain high levels of contaminants. If this is the case, the nitrogen source should be batched into the medium last to allow the minimum time for growth of the contaminant before sterilization is initiated.

3.2.3. *Additives*

(i)   *Vitamins.* These are a group of structurally unrelated compounds that are required by organisms in order to carry out normal metabolism. They are usually required in small amounts $(1-50 \ \mu\mathrm{g} \ \mathrm{litre}^{-1})$ and may either be supplied pure, or else as part of a complex medium. For instance many of the organic nitrogen sources mentioned above contain vitamins, and by careful blending the vitamin requirements for the medium may be met.

Common vitamins include thiamine, biotin, nicotinic acid, lipoic acid, and pantothenic acid, because of their complex structure they tend to be heat-labile. Two possible solutions to this are either to take account of destruction of the vitamin by increasing the level in the medium prior to heat treatment, or to use a different method (for instance, filtration), and sterilize the vitamins separately from the rest of the medium.

(ii) *Other additives.* Often other components are added to the medium to enhance productivity. These may include precursors of the desired product, for instance phenylacetic acid for Penicillin G production, or antibiotics to protect or positively select for the organism that is desired (this is increasingly common in animal/plant cell culture). Again many of these components will be heat-labile and it is therefore desirable not to use autoclaving or a similar procedure. Examples of other precursors include steroid precursors which are immiscible with water. Here, dissolution in a suitable organic solvent followed by filtration will yield a sterile solution which can be fed to the vessel following sterilization.

Included among additives are antifoams: agents that prevent the foaming associated with sparger aeration of complex protein-containing media. Foaming will cause loss of fermentation volume, introduce contamination risks, and may give mixing and oxygen transfer problems if a stable foam is produced. All antifoams are surfactants and are generally immiscible with water. The method of sterilization favoured is to produce an aqueous emulsion of the antifoam and sterilize this by heat. However, the problems previously mentioned in heat-treating water-immiscible substances should be noted. Filtration is difficult because antifoams are often polymers and are highly viscous in character.

Liquid filtration as a means of sterilizing media has become increasingly popular, especially on the small scale, where heat sterilization is more difficult to control. Filtration provides a cheap and satisfactory method of sterilization (see references 8 and 9). However, some components may adsorb on to the filter, reducing nutritional quality. Essentially, a sterilizing filter must remove from the process stream all viable microorganisms. Typically the size of organism that must be removed is about $1-5$ μm for bacteria and yeast, but may be as small as 0.01 μm for viruses.

Filtration is most suitable for sterilizing heat-labile components (vitamins and other additives), and these may then be added to the main medium once it has cooled. Filtration should not be considered as a method for sterilizing media that contain solids and that are reasonably insensitive to heat degradation.

3.3 Continuous sterilization

If a continuous fermentation is to be carried out, a continuous supply of sterile medium to the fermenter will be required. There are basically two options for sterilizing this medium supply:

(i)   The medium is sterilized batchwise and stored in a (sterile) feed reservoir, in sufficient quantity to supply the needs of the fermentation, for a significant part of the period. This then has to be linked aseptically to the fermenter, and the medium pumped in. Throughout the fermentation, fresh medium feed reservoirs may need to replace the original one, again aseptically.

(ii) Alternatively, non-sterile medium in a reservoir may be continuously pumped through a sterilizer to the fermenter.

The latter is less common at the laboratory scale.

If continuous sterilization is used, it is common to employ a high temperature $-$ short time (HTST) regime, involving three basic steps, and often a fourth optional, step (1):

(i)   Rapid heating to a high temperature (130 to 150°C), using direct injection of high-pressure steam (80 to 100 p.s.i.g.) or in a heat exchanger.

(ii) Holding at the sterilization temperature, via passing the medium through a lagged holding section. The residence time is typically a few minutes, or even less than a minute, depending upon the needs of the sterilization, and is achieved by matching the volume of the holding section with the medium flow-rate required.

(iii) Rapid cooling to below 100°C, in a 'flash cooler' or in a heat exchanger.

(iv) Further cooling to fermentation temperature, exchanging heat with fresh, incoming medium, (in a pre-heater), and then with coolant.

On the laboratory scale, there are continuous sterilizers available for use with small pilot-scale fermenters: e.g. the Alfa Laval 'Sterimedia Mini', which will process at a rate as low as 10 to 30 litre h^{-1}. However, at this scale and lower, batch sterilization is still the most commonly used method.

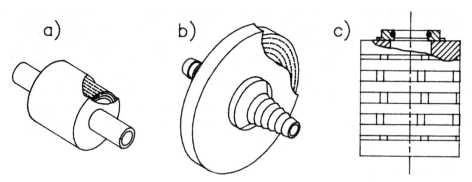

Figure 6. Types of air filter: (**a**) depth filter; cartridge form; (**b**) membrane (absolute) filter; in-line disc form. (**c**) membrane (absolute) filter, cartridge form.

4. STERILIZATION OF GASES

In fermentation processes the requirement for sterile gas will be high; a gas flow rate of 1 volume of air/vessel volume/minute is not unusual. Heat is not a viable option for sterilization of air because of the low thermal conductivity of air, and also dry heat is much less effective in destroying microorganisms than wet heat. For these reasons filtration is the only practical method for sterilizing gases.

The most common gas to be filtered will be atmospheric air, but other gases may be required in some fermentations, for instance oxygen, carbon dioxide and nitrogen. The gases are all treated in a similar fashion for sterilization. However, it should be noted that any lines and valves that are likely to come into contact with pure oxygen (or concentrations of oxygen significantly above atmospheric levels) should be entirely free from oil or grease deposits—such a combination is explosive.

Air must be supplied to an aerobic fermentation continuously for $2-14$ days for typical batch fermentations, and for longer periods in continuous operation. During this time all contaminating organisms must be removed from the air stream. Clearly the duty for a filter is quite exacting when it is considered that air from the compressor may contain up to 2000 microbes m^{-3}, and that exit air from the vessel will contain an even higher microbial concentration.

Air leaving the fermenter also needs to be sterilized to prevent the ingress of organisms through the air exit line and to prevent aerosols of organisms from the fermentation escaping into the working environment.

4.1 Selection criteria for filters

The primary criterion for filter selection is the ability of the filter to remove microorganisms from the air stream. Two types of filter are recognized: depth filters and absolute filters (see references 8 and 9).

4.1.1 *Depth filters*

These contain fibrous material, such as cotton, glass fibre, or steel wool. This may be packed into a glass or stainless steel housing attached to the appropriate air line. Alternatively, the fibres may be formed into a cartridge, such as the Balston resin-bonded glass microfibres, which are in the form of a porous tube (see *Figure 6(a)*). The cartridge is then fitted within an in-line sterilizable housing (most commonly of stainless steel) or sealed within its own in-line housing (such as a sterilizable polycarbonate housing). The distance between fibres is typically $0.5-15$ μm, compared with a size range of $0.2-1$ μm or larger for bacteria and spores. Consequently, there is a possibility that some particles may penetrate the filter. The greater the depth of the filter the lower the chance of an organism penetrating it. Richards (5) has defined a parameter, X_{90}, the depth of a filter required to remove 90% of particles. However, the X_{90} value is also dependent upon the air velocity.

4.1.2 Absolute filters

These are typically made from a hydrophobic membrane, such as PTFE. They have a fixed sub-micron structure of uniform size distribution, and are theoretically 100% efficient in removing microorganisms. The most commonly-used pore size for sterilization is 0.2 μm. However, some filters will retain much smaller particles. Absolute filters (or membrane filters) for laboratory and small pilot-scale applications come in a number of forms. Two common designs are shown in *Figure 6(b)* and *(c)*. Disc membrane filters are frequently used on the laboratory scale and as vent filters on addition vessels. Although they can be supplied separately and fitted to sterilizable stainless steel housings, the most common configuration is sealed within a sterilizable plastic in-line housing, as illustrated in *Figure 6(b)*, for example, filters sold by Gelman Sciences and Millipore. Hose tails on both sides allow rapid connection of flexible rubber tubing. To minimize the effects of accidental wetting, these filters are best mounted with the disc horizontal and air flow upwards. On the pilot scale, it is more common to use membrane cartridge filters, such as the pleated membrane Emflon cartridges of Pall and those manufactured by Domnick Hunter and Gelman Sciences. These are plug-fitted within steam-sterilizable filter housings connected within *in situ*-steamed fermentation systems. If choosing filters for equipment that is to be sterilized *in situ*, it is important that the housing is rated for this duty, too. Also some cartridges are designed for steaming in one direction only, because the support matrix softens during steaming and offers reduced support. It is therefore important that such units are so arranged that steaming is in the correct direction.

Depth filters tend to be hydrophilic, and their efficiency for removing particles is greatly impaired if they become wetted. Furthermore, organic material will become captured in the filter matrix, making it difficult to sterilize and producing conditions favourable to microbial growth. Absolute filters are more resistant to wetting. However, if it occurs, filtration efficiency may be reduced and pressure drops will be higher, leading to reduced air flows and, ultimately, to blockage. This will be accelerated if the wetting is caused by foam carried over from the fermenter. Overall, gas filtration efficiency is affected by the quality of the gas to be filtered. It is often advisable to pre-filter inlet air, to remove dust particles and droplets of oil and water, before it reaches the inlet sterilizing filter. This will also prolong the life of the filter. If gases are supplied from compressed-gas cylinders, it is not normally necessary to have them pre-filtered, provided that the gas lines used are clean. The moisture content of inlet air is not generally a problem, provided any water droplets have been removed. However, the air from the fermenter will be saturated with water vapour at the fermentation temperature, and may therefore wet the exit filter. Furthermore, droplets of medium could be entrained in this air stream, and may add to wetting problems. An added precaution, especially on pilot-scale fermenters, is to fit a catchpot between the condenser and exit air filter, to remove entrained droplets.

Exhausted depth filters are likely to allow significant numbers of organisms to penetrate the filter unit, whereas one would expect exhausted absolute filters to block. However, penetration of old absolute filters can occur. Therefore, it is advisable to discard a filter (depth or absolute) sterilized more than a certain maximum number of times, or to use it for no more than a maximum total period. Manufacturers should be consulted over this. An exception is the stainless steel filter, which may be regenerated. However, the manufacturer's procedure for this should be followed. As a rule, membrane-type absolute filters are more expensive than depth filters, so this extra cost may have to be balanced against the greater security absolute filters can afford. Nevertheless, it must be remembered that removal efficiencies for fresh depth filters of the correct grade are very high, too: For example, Balston quote a figure of greater than 99.9999% for their Grade SA filters.

Sizing a filter is important also, for too small a filter area will not only impede efficiency and reduce working life, but produce an unnecessarily high pressure drop in the air line. Increasing the size of filters used must be balanced against cost, so manufacturers' literature should be consulted in order to get the correct size for the air-flow duty required.

5. ASEPTIC OPERATIONS AND CONTAINMENT

Most sterilization operations take place before the fermentation begins. However, great care must be taken to ensure that no contaminant is allowed to invade the fermenter or its aseptic lines and ancillary components while the process is going on. Therefore, those parts of the system which are aseptic must be contained. This is done partly by equipment design and partly by maintaining 'barriers' to contaminants within fluid lines and around certain peripheral areas. The filtration of inlet and exit air represents one type of 'barrier', and has already been considered in this chapter. Other 'barriers' may also be needed, and examples of these are briefly discussed below. In addition to containment, measures must also be taken to avoid contamination when material is transferred into or out of the fermenter and other areas (such as inoculation and sampling), activities which often involve the 'making and breaking' of lines. Therefore, good aseptic design and practice is required, too. These details often vary with scale.

5.1 Methods used at different scales

Even within the area of laboratory fermentations, there are marked differences in the means used to prevent contamination during fermentation, in particular regarding the design and operation of inoculum lines/ports, transfer and harvesting lines, and sampling systems. These differences are often dependent upon the size of the fermenter, but most likely the principal influence is the material of construction of the fermenter. Vessels 10 litres or smaller are usually constructed out of glass, with stainless steel top/bottom plates; vessels 20 litres or larger are generally built in stainless steel.

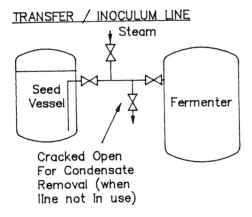

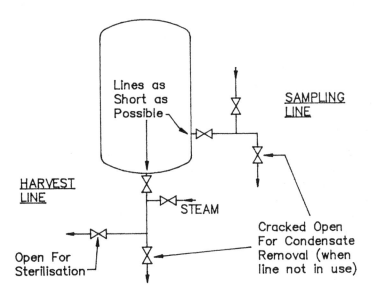

Figure 7. Line arrangements for aseptic operation.

5.1.1 *Lines*

Systems using autoclavable fermenters will normally be fitted with flexible, autoclavable lines such as silicone rubber tubing. The characteristics of various tubing types are discussed in Chapter 1.

Clamps and connections available for tubing have also been described in Chapter 1. However, it is worth mentioning here that all tubing should be securely fastened to connectors, vessels, etc., to avoid disconnection by a slight pressure build-up, such as that which could occur during cooling down in an autoclave. In doing this, it should be remembered that silicone rubber softens slightly while at sterilization temperatures, and can easily split if unduly stretched. Therefore the size of tubing and connector should be properly matched, to keep stretching to a minimum. Any rough edges should also be avoided on connections, hosetails, etc.

Most lines associated with *in-situ* sterilizable fermenters are usually rigid and constructed out of stainless steel, although flexible lines of braided stainless steel are often used where frequent making and breaking of lines is necessary e.g. for sampling or changing addition vessels. Both types of line are sterilized by steaming out and holding at the sterilization temperature, at the same time as the fermenter is sterilized. Other associated components such as air filter units and *in-situ* sterilizable addition vessels are, of course, steamed out too, at this stage. In addition to this, it is common practice to maintain in lines that are not being used at the time a steam 'seal' or barrier: steam at, say 121°C (15 p.s.i.g.) is passed into the section, with condensate leaving via a steam trap. (see *Figure 7*). This ensures a sterile 'barrier' between the vessel and the outside world, and also that any potential contaminant that may have been introduced on using the line is removed. Commonly, sample, transfer, and harvest lines, and those servicing steam jackets around pH and dissolved oxygen probes, are treated this way.

5.1.2 *Ancillary vessels*

These include addition vessels for medium, acid, alkali, and antifoam, product receiving vessels, and sample vessels, and sample vessels for larger fermenters. Whether sterilized by autoclaving or *in situ*, all vessels should be fitted with vent air filters at the top, to filter air admitted or expelled by liquid volume changes. These filters should be heat-sterilized with the vessels. Exceptionally, filters that have been pre-sterilized by the manufacturers (e.g. using gamma or ultraviolet radiation) may be aseptically fitted to the vessel after sterilization, provided the connecting line is sterile, plugged, and capped with foil, etc. Finally, any bungs fitted, say to aspirators, should be firmly fixed, possibly by wiring on, to avoid being dislodged by pressure increases during autoclaving or in operation.

5.2 **Inoculation and transfers**

Smaller fermenters are usually inoculated using shake-flask cultures. Unless each shake flask has been supplied with a side-arm, it is common to transfer its contents into an inoculum vessel prior to inoculation. This vessel will usually be fitted with a discharge arm at the bottom, connected to the fermenter via flexible tubing, and a vent filter. It is often autoclaved connected to the fermenter. In such cases, the line between the two must be clamped during sterilization, to prevent escape of fermenter medium. Alternatively the inoculum vessel may be sterilized separately, along with its connecting line, which is aseptically connected to the fermenter prior to inoculation.

Inoculation may then be performed by gravity feed (holding the inoculum vessel higher than the fermenter) or pumping using a peristaltic pump in the line. Any pressurization in the fermenter may have to be reduced first, especially with gravity feeding. After inoculation, the inoculum line must be clamped, and is usually removed from the

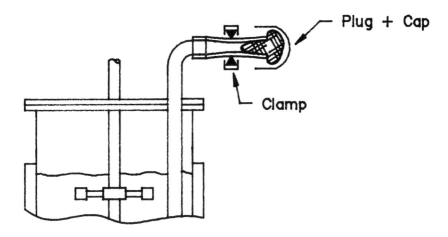

Figure 8. Simple sampling line.

fermenter. The inoculum port must then be aseptically blanked off. During inoculation, the exit air line must, of course, remain open.

Larger fermenters will generally be inoculated from a seed fermenter which is aseptically coupled for the operation, and then removed. Where rigid/braided steel lines are involved, when the lines are coupled, the transfer line is steamed out for several minutes before inoculation, and again afterwards, following which the lines are disconnected (see *Figure 7*). If the transfer is to be made through a silicone rubber line connected to a steamable inoculum port, it is essential to shut off the steam supply to the port before the inoculum line is connected. The following sequence is suggested:

(i)   Shut off the steam supply to the inoculum port.
(ii)   Remove the blanking cap from the inoculum port line, and aseptically fit the inoculum line. Connect this line to the peristaltic pump.
(iii)   Reduce the pressure in the fermenter. Ensure the exit air line is at least partially open. Unclamp the inoculum line.
(iv)   Open the inoculum valve on the fermenter and inoculate.
(v)   Close the inoculum valve on the fermenter. Disconnect the inoculum line, fit the clamp and remove the seed fermenter.
(vi)   Re-fit the blanking cap on the inoculum port line and turn on the steam supply to the inoculum port line.
(vii)   After a few minutes steaming, open the inoculum valve for a short time (say, 5−10 s), to clear the valve of old inoculum, then shut the inoculum valve. Maintain steam to the inoculum port line.

Similar procedures are followed at each scale for other one-off transfers and connections of fresh addition vessels. Note that in-line valves must be sterilized, too. Steaming-through is preferable, and is necessary for ball valves. However, diaphragm valves may be sterilized by steaming on only one side, if it is impractical to open the valve involved. This should influence the selection of types of valve used.

a) Simple sample hood

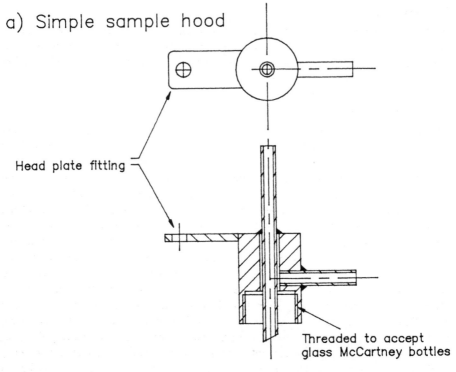

Head plate fitting

Threaded to accept
glass McCartney bottles

b) Sample Hood For In—Line Steaming

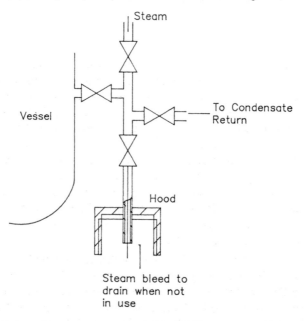

Steam

Vessel

To Condensate
Return

Hood

Steam bleed to
drain when not
in use

Figure 9. Sampling hoods: **(a)** simple; **(b)** for in-line steaming.

5.3 **Sampling**

A variety of sampling devices exists, and it is outside the scope of this chapter to survey them all. However, we shall illustrate methods of aseptic sampling procedure with reference to three sampling systems which it is hoped represent the range of sophistication available in laboratory fermenters.

5.3.1 *Simple sampling line*

This is illustrated in *Figure 8*. Typically, the silicone rubber tubing is clamped and the free end plugged when the fermenter is sterilized, and is left in that state until the first sample is required. At that stage, the plug is removed and the end of the tube is immediately immersed in ethanol. The procedure followed after this depends upon the sample receptacle used. For example, if a sterile syringe is used to draw off a sample, this is rapidly attached to the silicone tube, air to the vessel is shut off and excess head pressure allowed to reduce to prevent surging of sample into the syringe. The silicone tube is unclamped and sample drawn off. The tube is again clamped, excess sample downstream of the clamp allowed to drain rapidly, and the end of the tube immersed in ethanol until the next sample is taken. The vessel is re-aerated. Alternatively, the sample may be taken into a (flamed) open-necked bottle (e.g. a McCartney bottle). Here, a similar procedure may be followed as with the syringe, except that aeration is continued at all times. This is to maintain a positive pressure in the fermenter, both to force out sample and to reduce the likelihood of contamination entering the line.

Although this sampling arrangement is simple, it is also prone to contamination, either of the end of the silicone tube or via drainage back into the fermenter. Therefore, a system with more protection is recommended, especially for slow or sensitive fermentations.

5.3.2 *Sample hood*

Hoods of various designs are commonly used on fermenters from bench scale up to small pilot scale. Versions are available that are constructed in glass or stainless steel, and are in a variety of sizes depending upon the sample bottle required. Many are supplied with brackets for fitting to the fermenter top plate. A typical design is illustrated in *Figure 9(a)*. A sample bottle is screwed into the hood, and a filter cartridge fitted to the side-arm using a short length of silicone rubber tubing prior to sterilization of the fermenter. Sterilization is carried out with the bottle screwed loosely into the hood, to prevent stress in the glass due to differential thermal expansion. Also, the silicone rubber tubing connecting the hood to the fermenter sample tube is clamped, to prevent escape of medium into the bottle. After sterilization of the fermenter and cooling, the bottle is screwed more tightly into the hood, but the clamp on the line is kept shut.

Sampling may be done in at least three ways:

(i)   Aeration of the fermenter may be continued throughout sampling. Here, releasing the clamp on the connecting line allows sample to flow into the bottle, under the positive pressure in the fermenter. When a sufficient volume has been taken, the connecting tube is again clamped, the bottle unscrewed, and a fresh, sterile bottle fitted aseptically (usually with flaming of the hood and bottle neck). Undrained sample in line may be forced back into the vessel if aeration is

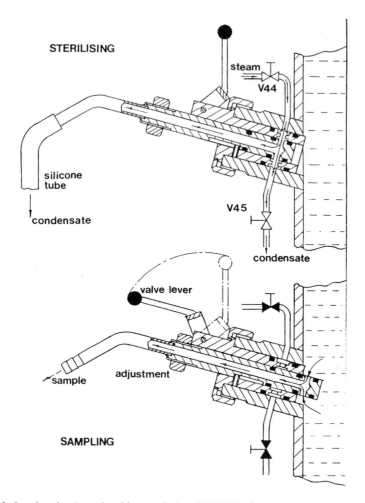

Figure 10. Sample valve (reproduced by permission of LSL Biolafitte).

temporarily stopped, the clamp on the connecting line released, and pressure applied through the side-arm filter, for example, via an empty syringe. A small amount of undrained sample may drop into the fresh bottle, but if the inlet hood tube is short, this will be a minimal volume. After draining, the connecting line is clamped and aeration re-started.

(ii)   Alternatively, aeration may be halted. Here, an empty syringe is fitted to the side-arm filter, the connecting tube clamp released, and the syringe used to draw sample into the bottle. Drainage is effected by reversing the air-flow in the syringe before clamping the line, replacing the bottle, and re-starting aeration.

(iii)   Hoods fitted to pilot-scale fermenters are likely to contain steam seals, as in *Figure 9(b)*. The hood is steamed through when the fermenter is sterilized, and

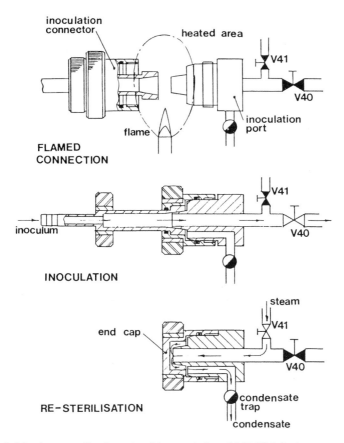

Figure 11. Quick-release coupling (reproduced by permission of LSL Biolafitte).

a small bleed of steam is maintained through the hood during fermentation. For sampling the following procedure has been used in our laboratories:

(a) Shut off steam to the hood. Clear the sample tube by steaming into vessel very briefly via the sample line. Then close the steam to the sample line.

(b) Allow the line to cool. Take a sample. Clear the sample tube by closing the line to the hood and briefly steaming through the sample tube to the vessel. Close the vessel line and maintain a bleed of steam through the hood.

5.3.3 *Sample valves*

These are less common in small bench-scale fermenters, but more common in stainless steel fermenters. There, they are supplied with steam seals for maximum protection against contamination. Once again many designs have been developed. *Figure 10* shows an LSL Biolafitte sample valve designed for rapid and simple operation. A silicone rubber or flexible stainless steel line connects the valve to a condensate return line for normal steam-bleeding, and is fitted for quick release before sampling.

(i)   During sterilization, the valve lever is fully retracted towards the vessel, allowing steam to flow through both the external valve jacket and the valve itself. After sterilization, the steam valve may be cracked open to allow a slow bleed of steam through the valve and its jacket. Alternatively, these may be steamed out prior to sampling.

(ii)   Sampling is effected by turning off the steam supply to the valve, allowing the valve to cool, then pulling the valve lever forwards until the required volume is collected. Following this, the lever is fully retracted and the valve and its jacket steamed out, as in (i).

5.4 Making and breaking lines

Again, there are a variety of ways of making and breaking lines in laboratory fermentation systems. These operations may vary in how frequently they must be performed. For example, an inoculation line will need to be made and broken before and after the operation. Addition vessels will need connection to the fermenter before inoculation, with further replacements possible if the fermentation is long, especially the feed vessel for a continuous fermentation. All of these operations carry the risk of contamination, so must be performed rapidly and with care.

At the simplest level, a silicone rubber line may be connected to another silicone rubber line or to a vessel via a stainless steel or glass tube. This may be fitted with hose tail ends for firmer grips (see also section 5.1). When a line is to be broken, the lines at both ends of the connector are clamped to prevent leakage, the line to be removed is taken from the hose tail/tube, and the latter is flamed. The fresh line (sterilized plugged and capped) is unplugged. Its end may first be dipped in ethanol and is then rapidly connected to the hose tail/tube. Finally the line clamps are removed. This system is simple and cheap, but has to be handled skilfully and rapidly to prevent contamination. Therefore, quick-release couplings are often preferred.

Several quick-release line couplings are available for aseptic use. Most designs use a hose tail for connection to a silicone rubber (or other flexible) line, and a male/female screwed end for the quick-release end. Such couplings are obviously more expensive than simple hose tails, but are easier to use and give greater safeguards against contamination (*Figure 11*).

6. DISPOSAL

One final area where sterilization is required is a less obvious one. At the end of a fermentation, a number of items will be left containing biological material, and, in particular, living organisms. These include fermenters containing cells and spent medium, sample bottles, pipette tips, Petri dishes, and other associated equipment. Furthermore, there will be full or empty addition vessels which would support microbial growth. All of these items will require the use of disposal procedures. It is not normally acceptable to dispose of significant quantities of biological material without rendering it harmless, or free from potentially or actually harmful contamination. Even when a non-pathogenic organism has been produced, release of significant quantities may cause allergic or microtoxic reactions in people. Also, if there has been any contamination from the atmosphere, this should be treated as potentially pathogenic. Therefore sterilization of materials prior to disposal, and of equipment prior to cleaning and further use, is a common requirement of many organisations. It is common practice to be guided by the recommendations of the relevant Advisory Committees. (For example, references 10 and 11.)

As a general guideline, fermenters and contents not required for further processing should be sterilized in the same way as for preparing for a fermentation, that is, by autoclaving or *in-situ* steaming. After sterilization, the (cooled) contents should be mixed with an acceptable disinfectant and flushed down the drain, avoiding splashing, with copious volumes of water. Glassware, etc., should be autoclaved. Materials for disposal (syringes, pipette tips, Petri dishes, etc.) may be autoclaved in unsealed containers, or placed in suitable sealed containers for disposal, usually by incineration. In addition, it may be a requirement that, for example, all liquid effluent from a continuous fermenter is pumped into a 'kill tank' and mixed with disinfectant before passing to the drain.

7. ACKNOWLEDGEMENTS

The authors wish to acknowledge the co-operation and guidance given by Professor A.W.Nienow and Mr A.N.Emery, and the financial support of SERC/DTI and Life Science Laboratories, Ltd., through a Teaching Company Scheme.

8. REFERENCES

1. Banks,G.T. (1979) In *Topics in Enzyme and Fermentation Biotechnology*, Wiseman,A. (ed.), Ellis Horwood, Chichester, Vol. 3, p.170.
2. Ayres,J.C., Mundt,J.O. and Sandine,W.E. (1980) *Microbiology of Foods*, W.H. Freeman, San Francisco.
3. Bailey,J.E. and Ollis,D.F. (1986) *Biochemical Engineering Fundamentals* (2nd edn), McGraw-Hill, New York.
4. Stanbury,P.F. and Whitaker,A. (1984) *Principles of Fermentation Technology*. Pergamon Press, Oxford.
5. Richards,J.W. (1968) *Introduction to Industrial Sterilisation*. Academic Press, London.
6. Russell,A.D. (1982) *The Destruction of Bacterial Spores*. Academic Press, London.
7. Moo-Young,M. (ed.) (1985) *Comprehensive Biotechnology*. Pergamon Press, Oxford, Vol. 2, Chapter 14.
8. Moo-Young,M. (ed.) (1985) *Comprehensive Biotechnology*. Pergamon Press, Oxford, Vol. 2, Chapter 17.
9. Matthew Hall Norcain Engineering, Ltd. (1984) *Process Plant for Biotechnology*, Biotechnology Unit of the Department of Trade and Industry, London, Study Numbers PPFB 1 and PPFB 2.
10. DHSS (1978) *Code of Practice for the Prevention of Infection in Clinical Laboratories and Post Mortem Rooms (the Howie Code)*. HMSO, London.
11. Advisory Committee on Dangerous Pathogens (1984) *Categorisation of Pathogens According to Hazard and Categories of Containment*. HMSO, London.

CHAPTER 9

Animal cell fermentation

MARK LAVERY

1. INTRODUCTION

In recent years, interest in the *in vitro* cultivation of animal cells has increased because of the need for production of interferon, monoclonal antibodies, various hormones, vaccines, and other cell products with potential human and veterinary applications. This has prompted research into basic biology and physiology of the cell which has seen the development of a variety of systems for the *in vitro* growth of animal cells. Some are a result of modifications or extensions to microbial stirred tank reactors (STR). Others have seen the emergence of air-lift reactors for submerged fermentation of animal cells. Further developments have involved intensification of the process, allowing fermentations to be operated at high cell densities. Newcomers to animal cell fermentation may find it difficult to select a system from the vast range available. It is not the intention of this chapter to recommend one manufacturer's system over another but rather to describe the operation of those that are currently available and to guide the beginner towards successful animal cell fermentation.

2. THE NATURE OF ANIMAL CELLS

Animal cells can be classified into two groups according to their different modes of growth.

(i) *Anchorage-dependent cells* require a solid substratum (support) for their replication. They produce cellular protrusions called pseudopodia or lamellepodia which allows them to adhere to positively charged surfaces. Another important property of this cell type is the phenomenon of contact inhibition and monolayer growth. Non-transformed cells such as primary and diploid cell lines fall into this category.

(ii) *Anchorage-independent cells* do not require a support and can be grown as suspension cells in submerged culture. Established and transformed cell lines generally fall into this category. Cells of an intermediate type also exist. Baby hamster kidney cells (BHK) used typically in the commercial production of foot and mouth disease (FMD) vaccine can be grown as either anchorage-dependent or suspension cells. By nature they are anchorage-dependent cells. However, an established line, BHK-21 cells, is cultured industrially as suspension cells in fermenters of 1000–5000 litre volume for FMD vaccines.

Both cell types lend themselves to growth in submerged culture using standard laboratory fermenters, but with the anchorage-dependent cells one of the main problems is providing a surface for attachment and still maintaining a suspension state. This was overcome by the development of microcarrier cultures of animal cells by Anthony Van Wezel in 1967. Cells have a net negative charge and will form an adhesion to a positively charged surface by electrostatic forces. Van Wezel exploited this property by allowing the cells to attach to DEAE−Sephadex A-50 beads. Initial work was affected by charge toxicity although the development of low-charge DEAE−Sephadex beads overcame the problem. The cells grow and form a confluent layer on the surface and can therefore be cultured in traditional submerged culture as quasi-suspension cells.

Animal cell lines with transformed properties such as hybridomas and established lines which have been adapted to grow as suspension cells, e.g. BHK-21, can be cultured in a homogeneous submerged condition and therefore resemble the more traditional microbial fermentation. However, there are fundamental differences between microbial (prokaryotes) and animal (eukaryotes) cells which require different vessel design and culture strategies to be adopted when growing animal cells in standard microbial fermentation systems.

3. DIFFERENCES BETWEEN MICROBIAL AND ANIMAL CELLS

Basic differences in cell structure and nutritional requirements between microbial and animal cells dictate the use of modified vessels and the development of a more gentle gassing and agitation regime for cultivation of the latter. However, to say it is more difficult to grow animal cells *in vitro* than it is to culture microbial cells would be quite wrong. Some microorganisms have nutritional and cultural requirements which can be just as complex as animal cells.

In general, however, due to the differences listed in *Table I*, animal cells do require modified culture strategies for growth in standard microbial fermenters.

More care and attention to detail should be a prerequisite for the fermentation stage and all the preceding stages. With this in mind, fermenter equipment manufacturers tend to supply either modified microbial systems for animal culture or dual-purpose bioreactors. Others such as Setric Genie Industrial (SGI) will provide fermentation systems which have been designed and constructed specifically for animal cells. As a rule most manufacturers will supply systems built to customers' own specifications. A list of equipment manufacturers is included in the appendix (p. 221).

Table 1. Generalized differences between animal and microbial cells.

Features	Microbial cells	Animal cells
Cell wall	Generally present	Generally absent
Cell membrane	Present	Present
Growth rate	$0.1-0.5 \text{ h}^{-1}$	$0.01-0.05 \text{ h}^{-1}$
Oxygen requirements	High	Low
Nutritional requirements	Usually simple	Complex
CO_2 requirements	Some microbes require for growth	Forms part of buffering system in most cell culture media
Environmental effects	less affected	highly susceptible
Size range	$100-2000$ nm	$10\,000-100\,000$ nm
Seeding density	One cell (theoretically)	10^5 cells ml^{-1}
Growth density	High (10^9-10^{10} cells ml^{-1})	Low (10^6 cells ml)

4. MODIFICATIONS TO STANDARD STIRRED TANK REACTOR

A standard microbial laboratory stirred tank reactor (STR) consists of a flat-bottomed vessel usually constructed of Pyrex glass, and a stainless steel top plate. The top plate has ports drilled which allow access to the vessel for heater and temperature probes, sensors for oxygen and pH measurement and a mechanical seal through which a shaft enters to provide agitation of the medium via disc impellers. Gassing is accomplished by sparging air through a tube which is situated just below the bottom impeller. This type of vessel would be unsuitable for animal cell fermentation as it stands, and would require modification for successful growth of animal cells to be achieved. During the culture of animal cells in fermenters physical factors such as agitation, pO_2, pCO_2, and pH are all interrelated and can have either individual or combined effects on the various cell growth parameters as illustrated in *Figure 1*. This in turn will have an effect on the culture strategy and the design of the reactor. It is therefore essential to bear in mind that any modification made to one aspect of the fermenter will have effects on the others. Additionally, product accumulation and removal also affect culture regimes and reactor design.

4.1 **Agitation of cells in culture**

One of the most important criteria when choosing a system for animal cell fermentation is the method of providing suspension of the cells. Agitation has two functions. It assists the mass transfer between the different phases (gas, liquid, and solid) present in the culture, and it mixes so maintaining a homogeneous chemical and physical environment. This is particularly important for heat transfer and for mixing during pH control and substrate feed. Animal cells generally do not possess a cell wall. The interface between the culture medium and the cytoplasm is therefore a thin cell membrane. Traditionally microbial fermenters are equipped with vaned disc-type impellers as illustrated in *Figure 2*. These produce very high horizontal shear forces which can damage animal cells; better mixing at similar speeds is achieved by using marine or pitched-blade impellers. They provide horizontal and vertical mixing and are generally less damaging to animal

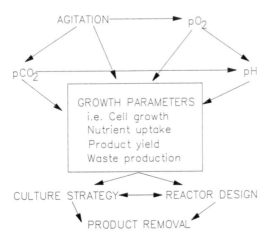

Figure 1. Interrelationship of physical conditions during animal cell culture in fermenter.

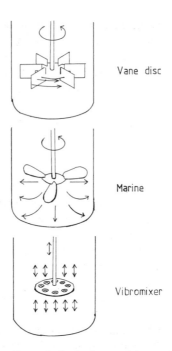

Vane disc

Marine

Vibromixer

Figure 2. Basic approaches to medium agitation in fermentation vessels.

cells. When vigorous mechanical agitation is employed in the absence of baffles, vortexing may occur. This can prove more damaging to the cells than any mechanical stress effect of impellers and should be avoided in animal cell culture. Standard laboratory microbial fermenters are generally manufactured with motors which will run uniformly between 100 and 1500 r.p.m. Although this speed range may be suitable for some animal cells it is more appropriate to employ a gear system (1:3) to reduce the revolutions. Good motors with appropriate gears will provide agitation as low as 10 r.p.m. Although the motor speed is often quoted as a specification it is more appropriate to consider the impeller tip speed, t_s, which is a function of motor speed and impeller diameter:

$$t_s = \text{r.p.m.} \times \text{impeller diameter} \times \pi$$

This is particularly important when moving from one scale of culture to another, e.g. when using a 1-litre fermenter to seed a 10-litre volume vessel.

Another form of vessel medium agitation which utilizes the Bernoulli effect is the vibromixer. This system has been employed for laboratory fermentations with animal cells. In addition to providing vertical mixing, it also promotes gas transfer and produces less shear and vortexing. A further advantage of this system is the absence of a mechanical seal which eliminates the possibility of contamination being introduced through the seal.

Magnetic agitation of a Teflon-coated impeller can also be employed to maintain cells in suspension in standard laboratory fermenters. In common with the vibromixer, the method also reduces the risk of contamination being introduced through a mechanical seal. It also provides a slow and gentle form of agitation more appropriate for animal cells. Magnetic agitation of culture in small flasks is an appropriate means of fermenter seed preparation and will be discussed later in this chapter.

A unique impeller design which also acts as an aeration device is a feature of the Celligen system manufactured by New Brunswick. The impeller design is said to provide efficient mixing and high oxygen transfer rates for animal cell fermentations. The nature of the impeller design is reported to generate very little mechanical shear and is therefore suited to shear-sensitive cells.

4.2 Methods of aeration

Aeration of a stirred tank reactor is required to provide cells with an adequate supply of oxygen in a form that can be assimilated. In animal cell fermentation, aeration agitation and pH control are all interrelated as previously illustrated in *Figure 1*. The role of aeration in pH control will be discussed later.

During standard microbial fermentations with oxygen-requiring organisms, air is sparged into the medium through a tube below the bottom impeller. High air-flow rates are used which results in bubbles being released into the medium and impinging on the impeller. Vaned disc impellers are very efficient at effecting oxygen transfer into the liquid phase.

The relative value of $k_L a$ (the volumetric O_2 transfer coefficient) and hence the oxygen transfer rate (OTR) will be affected by impeller speed, the air sparge rate and vessel internals (baffles). In addition, vessel geometry and impeller design also have critical effects on the oxygen transfer characteristics (Chapter 2).

Marine impellers are not as efficient in oxygen transfer as vaned disc types and in animal cell fermenters the low agitation speeds also produce low OTR. In microbial fermentation, $k_L a$ values of $5.4-44.2$ min^{-1} can be achieved, whereas with animal cell fermentations it is more common to encounter values in the range $0.017-0.42$ min^{-1} with sparged aeration and $0.0017-0.067$ min^{-1} with surface aeration. There is however a balance in that the low OTR experienced in animal cell fermentation is matched by the low oxygen requirements of animal cells. Oxygen demand of bacterial cells is in the range $0.43-2.33 \times 10^{-3}$ g cell^{-1} h^{-1}. Animal cells have oxygen uptake rates in the range $2-10 \times 10^{-12}$ g cell^{-1} h^{-1}.

The traditional method of supplying oxygen to cells is by bubbling through a sparge tube into the medium. Some animal cell types are susceptible to damage in this way from bubbles. To prevent this, headspace or surface aeration is employed. However, although the oxygen demand of animal cells is low the OTR via the surface is normally not sufficient to supply the cells and oxygen depletion can occur rapidly as cells grow. The $k_L a$ obtained using this method of aeration will be dependent on surface area. A larger surface area/volume ratio will provide better oxygen transfer. However, fermenter configuration limits the surface diameter, and, to overcome this, pure oxygen is used to enrich the air supply. Care should be taken when employing oxygen supplementation of air, since high oxygen concentrations can prove toxic to animal cells. The supplementation should be linked to an oxygen amplifier and controller to prevent this occurring.

Silicone tubing has also been employed as an aeration device in deep submerged culture of animal cells where bubble damage is apparent. Coils of thin-walled silicone tubing are submerged in the medium and air or oxygen-rich air is passed through it. Oxygen diffusion through the tubing wall and into the medium provides bubble-free oxygenation of the culture fluid. Braun manufacture a fermentation system employing this principle.

In cases where bubble damage to cells is not a problem, a small sinter tube can be fitted to the sparge line. This effectively does the work which vaned disc impellers do in microbial fermentations, i.e. producing small bubbles for oxygen transfer. The small bubbles can then be effectively dispersed by the propellers.

The condition of the gas supply for aeration of animal cells cultures is probably more crucial to successful culture than is the case with microbial fermentations. Oil and particulate-free sterile gas is required. If the air supply is from a standard commercial compressed-air generator, filters and traps should be installed in the process lines to remove oil and particles from the stream. For animal cell cultivation $5-10\%$ carbon dioxide is blended into the final gas stream. This is required principally for buffering, since cell culture medium employs a bicarbonate/carbon dioxide buffer. This will be discussed in detail later. Blending of carbon dioxide into the air can be carried out by a dual mass flow controller, but the initial outlay can be expensive. An alternative is to use standard commercial compressed cylinders of air with the desired concentration of carbon dioxide in air. The blend can then be passed through a cotton fibre bulb to remove any potential particulate matter, although in general the blend in commercially available cylinders is of good quality. Control of the gas flow from compressed cylinders can be achieved by using a simple rotameter calibrated from $0-100$ ml min^{-1}. Before introduction of the gas stream into the fermenter it should be filtered using hydrophobic-type vent filters with $0.1-0.2$ μm rating. Similar filters are required on the air outlet line with a condenser to reduce moisture build-up on the filter.

4.3 pH control

Traditionally cell culture media have been buffered with a bicarbonate/carbon dioxide system. The normal starting pH range of the most commonly employed tissue culture media is between 7.00 and 7.80. The optimum for each cell type will vary but it is usually within the physiological range. Since the pK_a for bicarbonate is 6.1, buffering in most animal cell culture medium occurs outwith the suboptimal range. Phosphate buffer has also been employed; however, the levels required for effective buffering are generally toxic to animal cells. Biological buffers such as Hepes are often employed in addition to the bicarbonate/carbon dioxide system. The pK_a for Hepes is 7.0, which makes it an effective buffer in the range $6.8-7.2$. Hepes is generally toxic to most animal cells above 25 mM; however, it is better to assess any toxic effects on individual cells. Phenol red is generally used in animal cell culture medium as a visual indicator of the pH value. It may prove toxic at high levels, but in reduced concentrations, and with experience, it can act as a very useful indication of the need for medium changes. *Table 2* indicates the amino acids, vitamins and minerals which make up a standard tissue culture medium. The quantity of each constituent will vary depending on the medium. The most common media in use today are RPMI 1640, Dulbecco's Modified Eagle's Medium (DMEM), and modifications of those standard formulations.

Several methods have been employed to control the pH during animal cell cultivation in fermenters. Separate supplies of carbon dioxide and air can be injected into the medium via solenoid valves. The operation of these valves is triggered by the pH controller.

Table 2. Typical composition of animal cell culture medium, e.g. Dulbecco's Modified Eagle's Medium (DMEM).

Arginine	Ca panthothenate
Cystine	Choline
Glutamine	Folic acid
Glycine	Inositol
Histidine	Nicotinamide
Isoleucine	Pyridoxal
Leucine	Riboflavin
Methionine	Thiamin
Phenylalanine	
Serine	$CaCl_2$
Threonine	$Fe(NO_3)_3$
Tryptophan	KCl
Tyrosine	$MgSO_4$
Valine	NaCl
	$NaHCO_3$
Glucose	NaH_2PO_4
Phenol red	
Sodium pyruvate	

Medium gassed with 5% CO_2 in air

(Animal sera usually added for growth of cells)

Alternatively carbon dioxide can be blended into the main stream air supply using a dual mass flow controller. The amount of carbon dioxide in the blend is influenced by this and low signals from the pH controller. As the pH drops, less carbon dioxide and more air is injected. In addition to this a sodium bicarbonate solution (about 10%) can be used to prevent the pH from decreasing below an optimum level. As an alternative to bicarbonate additions, 0.1 M sodium hydroxide and 0.1 M hydrochloric acid can be employed as in traditional microbial fermentations at laboratory scale. However, because of the low agitation and poor mixing during animal cell cultivation, localized areas of intense acidity or alkalinity may develop. If employing this method it is advisable to situate an impeller (or propeller) just below the medium surface to aid in titrant dispersion.

Carbon dioxide was thought to be a metabolic requirement for animal cell growth, but the evidence available does not support this. Some cell types have been successfully cultured without using carbon dioxide in the main stream gas. Animal cells produce carbon dioxide during culture which may be sufficient to meet any metabolic demand for the gas. On the other hand bicarbonate does appear to have some metabolic function and is required as a medium constituent. Buffering capacity in addition to bicarbonate/carbon dioxide can be provided by adding tris/citrate to the medium.

4.4 Vessel design

The method chosen for aerating an animal cell fermentation serves a dual function of providing oxygen necessary for cell growth and facilitating pH control. Factors such as gas flow rate, supply of gas (headspace or sparge), bubble size, agitation, gas composition and culture surface area are all important in controlling dissolved oxygen levels and pH of the medium. Of critical importance to this is vessel geometry. Whereas traditional microbial fermentation vessels have flat bottoms, those used in animal cell culture are either of a domed-base configuration with straight walls, or a round-bottomed flask. A domed base ensures that stagnant areas of poor mixing are unlikely to occur, particularly at low speeds. Baffles can be included in the type of vessel design to aid mixing and prevent vortex formation. On the other hand, round-bottomed flasks provide a larger surface area/volume ratio which would promote better gas transfer and control by surface aeration. Since this type of flange vessel is not generally available commercially, it normally requires the services of a glassblower to construct such a vessel.

4.5 Temperature control

Temperature control of vessels for laboratory microbial fermentations can be accomplished by the use of a cartridge heating element immersed in the culture medium. Providing the output from the element is of a sufficiently low wattage, it may be applied to animal cell fermentations at the $1-2$ litre scale. At larger scales, the output required to maintain the medium at constant temperature would adversely affect the cells. The most efficient method of temperature control for animal cell fermentation is a water-jacketed vessel. Water just above the desired temperature is heated up in a standard laboratory waterbath equipped with a pump. The water is then circulated through the jackets to maintain the temperature. Standard units can be supplied by manufacturers to perform a similar function. One of the problems associated with using a water jacket is that after some time the jacket gets rather grubby and requires cleaning. This can be eliminated by circulation of hot air instead of water through the jacket to maintain temperature at a constant level. A complete fermentation system which includes this method of temperature control is available commercially from Setric Genie Industrial (SGI).

Heater tapes or pads attached to a non-jacketed vessel are also suitable for temperature control of vessel medium on the laboratory scale. A good temperature controller with the correct setting of the proportional, integral and derivative terms (PID) will give very accurate and stable temperature control during animal cell culture.

5. AIR-LIFT REACTORS

Although it should be possible to culture most animal cells *in vitro*, not all will be amenable to growth in STR, even with the modifications and precautions discussed. In this case, the air-lift reactor can be used on the laboratory scale to produce the desired product. The principle of this type of fermenter is illustrated in *Figure 3*. The air/carbon dioxide mixture is sparged into the reactor at the base. As it rises, liquid at the top becomes oxygen-rich and hence more dense. It begins to fall, replacing the less dense medium at the base, promoting liquid motion within the vessel. However, mixing efficiency is lower than in conventional STRs and excessive foaming can occur. This can be overcome by extending the draught tube to just below the surface and just above the base of the vessel. A major advantage of this type of reactor for animal cells is the reduced shear forces compared with those generated in mechanically stirred tanks.

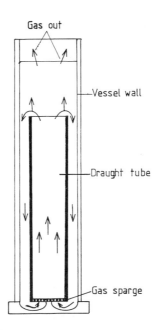

Figure 3. Principle of the air-lift fermenter.

This enables the cells to be cultured under a gentler mixing regime. Another advantage is the absence of moving parts and hence of a mechanical seal, which can be a potential area of contamination. The medium temperature can be controlled by circulating water through a jacket, and ports can be made available for the on-line measurement of the process variables. The generation of small bubbles at the base and increased gas hold-up results in very efficient OTRs in these vessels.

An extension of this approach to animal cell fermentation would be to employ a vessel with the air-lift principle and an impeller at the base of the draught tube. This combined vessel would provide high OTRs and good mixing in a single reactor. The air-lift reactor is used extensively for fermentation with animal cells and is employed commercially, by Celltech amongst others, for the production of monoclonal antibodies.

6. HIGH-DENSITY FERMENTATION WITH ANIMAL CELLS

Batch fermentation with animal cells of the anchorage-dependent or suspension type in submerged culture using STR or air-lift reactors is normally characterized by low cell densities. During fermentation with microbial cells, operating densities in the range $10^9 - 10^{10}$ cells ml^{-1} can be accomplished. In contrast, because of the sensitivity of animal cells to environmental changes, maximum cell densities over 3×10^6 cells ml^{-1} are usually not obtained. While this may be adequate for virus production systems, it certainly has shortcomings for product yields with secreted proteins such as monoclonal antibodies. Another disadvantage of such low product yields for proteins emerges in the bioprocessing and recovery stages, where small levels of the desired product have to be isolated from the other components, particularly serum proteins, present in the medium. It would, therefore, aid in bioprocessing if the material could be produced at higher levels per unit volume of medium and with less contaminating proteins. This is the purpose of the process-intensified systems which have been developed over the past two years. Some use existing STR and air-lift technology and others are a result of new bioreactor developments.

6.1 Perfusion culture

Stirred tanks and air-lift fermenters can be operated in this mode. It is generally achieved by withdrawing medium through a stainless steel or ceramic sinter. This mode of culture is often referred to as spin filter, since the sinter is rotated to prevent blocking from the cell mass. The rotating filter generates a laminar liquid layer beneath the surface of the filter. The centrifugal forces generated by the rotation force the cells to move away from the filter and maintain the liquid layer at the filter surface cell-free, hence preventing blocking. Spent medium can therefore be withdrawn, leaving the cells in the reactor, and fresh medium can be added to replenish utilized nutrients while also diluting out toxic products which may prevent cell growth or product secretion. Cell densities in the region of 3×10^7 cells ml^{-1} have been reported with this method. One of the problems associated with high cell density is the increased demand for oxygen, which has to be satisfied by using pure oxygen. However, the requirement for serum is reduced at high cell density culture and since better product generation can be achieved, the isolation process may prove to be easier. Perfusion culture is particularly applicable to microcarrier cell culture but, with sinters of suitable pore, this technique can also be used with suspension cells, where secreted protein can be continually removed and purified from the perfusate. The pioneering work of Feder and Tolbert has led to the commercial application of perfusion-type culture for animal cells by Monsanto and Invitron in the USA.

6.2 Encapsulation

This approach to high-density culture of animal cells can take two forms. The first is a kind of microencapsulation where the cells are retained within a membrane with small pore size. Nutrients and waste products pass freely through the membranes, and the cells and high molecular-weight secreted products are held inside. This technique has been exploited commercially for the production of monoclonal antibodies. The Encapcel technology of Damon Biotechnology employs polylysine beads $50-500$ μm in diameter. Densities in the region of 5×10^8 cells ml^{-1} and product levels 100 times higher than in conventional *in vitro* culture have been reported. The cell-containing capsules can be cultured in conventional animal cell reactors and are afforded protection from mechanical stress by the enclosing membrane polymer. Although the high cell density increases oxygen requirements, serum levels can be drastically reduced and the product may be obtained at a higher concentration and in a purer state.

A second method of immobilizing cells which can then be cultured in standard animal cell fermenters involves entrapment of the cells within an alginate, agarose, or fibre matrix. The cells grow freely within the matrix and secrete products into the surrounding medium. Cell densities in the region of 5×10^7 cells ml^{-1}, and higher productivity than conventional batch systems, have been reported.

A similar concept to free cell immobilization, which does not involve the use of STR or air-lift reactors, is inertially entrapping the cells in a porous ceramic matrix. Medium is continually pumped through the unit and the product can be collected in the waste medium. Alginate matrix immobilization (Karyon Technology) and ceramic matrices (Opticell, Corning) have both been used commercially for the production of animal cell products by fermentation.

6.3 Membrane-based reactors

The most popular type of membrane system in use for animal cell fermentation is the hollow fibre reactor. This consists of a bundle of hollow fibres held together within a cartridge, and operates on the principle of capillary action of the fibre. Medium is perfused through the fibres (tube side) and cells grow in between the fibre bundle (shell side). With suitable environmental and nutritional conditions, densities of 4×10^8 cells ml^{-1} can be obtained. The molecular weight cut-off of the membrane can be selected either to retain the product within the shell side or release it into the perfusing medium. This type of reactor is well suited to shear-sensitive cells, and in addition it provides a high yield and concentration of secreted product. As is common with high-cell-density culture of animal cells, serum requirements are reduced and it is usually relatively easy to replace the serum requirements with purified serum proteins. The hollow fibre reactor is the basis of the Endotronic system for animal cell products.

Other membrane-based systems for intensification of animal cell processes are commercially available. The Membroferm culture unit of MBR Bioreactor Ltd utilizes semi-permeable membranes to separate regions of high cell density from liquid bulk flow. However, instead of a tubular-type construction, flat sheets of membranes separated by fluorohydrocarbon nets are employed. This forms chambers of $15-20$ ml in volume, and typically 60 such chambers form a single unit. Again the molecular weight cut-off can be configured to suit a particular process. Another type of system commercially available is the microprocessor-controlled bioreactor from Bioreactor Technology Inc., the heart of which is a four-chambered stainless steel reactor. The reactor chambers have different volumes and are separated by packets of membranes. The central compartment 2 is a mechanically stirred reactor in which the cells are cultured. Chamber 1 is used to diffuse medium under pressure into the growth chamber. Chambers 3 and 4 are used to purify and concentrate secreted products. Yet a further innovation in membrane reactors has recently been announced by Bioengineering Limited. It consists of a STR within a larger STR separated by a cuprophan cellulose membrane of 10 000 molecular weight cut-off. Both have independent bottom-drive mechanical agitation with the ability to install bioprobes for separate monitoring and control of each reactor. The potential for a well-mixed, homogeneous, high-cell-density animal cell reactor is provided by this system.

7. CELL MAINTENANCE AND SEED PREPARATION

One of the most crucial areas of animal cell fermentation is the maintenance of cell stocks and seed preparation. In contrast to microbial cells (Chapter 3), animal cells do not survive lyophilization and usually do not survive storage at temperatures above

$-25\,°C$. Even between $-70\,°C$ and $-140\,°C$ physicochemical changes occur. The most suitable system for storage of animal cell stocks is liquid nitrogen at $-190\,°C$. However, since it is not possible to plunge the cells directly into liquid nitrogen, a controlled gradual decrease in temperature, the rate of which will depend on the cell type, is usually carried out. Although it may not be applicable to all cell types, a glycerol bath at $-20\,°C$ is an inexpensive way of achieving a gradual temperature decrease over a 30 min period. With important cell lines it is crucial to maintain a liquid nitrogen backup system in case of mishaps with the first storage container. When new cells are required they should be removed from liquid nitrogen and rewarmed as quickly as possible. For maximum survival rate they should be placed immediately in a $37\,°C$ waterbath to promote rapid thawing. It is important to have this stage in the process optimized for the particular cell type in use. Any loss of cell viability at this stage means that the recovery or resuscitation process will take much longer. When the cells have been thawed from frozen it is essential to place them immediately into pre-warmed $(37\,°C)$ suitable tissue culture medium at the desired pH and placed into a standard carbon dioxide incubator. In addition to the freezing and thawing process, the viability and general condition of growing cell stocks maintained in an incubator is also of crucial importance to successful animal cell fermentation. Several methods are available for assessing cell viability; the most common method is the use of trypan blue. Generally one drop of a 1% solution to nine drops of culture is used to assess the population's viability. Using this method dead cells which are unable to metabolize the dye stain blue, and live cells remain clear. However, the art of the animal cell technologist is to be able to assess cell conditions visually under an inverted microscope. This comes with experience and knowledge of the limitations of the particular cell line in use. *Figure 4* shows scanning electron microscope pictures of NS1-derived mouse hybridoma cells at different stages of culture in static tissue culture flasks. At day 1 the cells are $98-100\%$ viable and have a rounded and uniform appearance. By day 3 the cells are approaching the stationary phase of growth and beginning to show signs of viability loss. When day 5 is reached large sections of the cell membranes appear to be missing as the cell viability has been reduced to $10-20\%$. At this stage it would prove very difficult if not impossible to initiate cell recovery and a return to 100% viability. It is therefore important to carry out a split culture regime before this stage is reached. However, care has to be taken since animal cells require a minimum seeding density for successful growth. The seeding level will vary with cell type, but it is typically in the range 10^4-10^5 cells ml^{-1}. Therefore to initiate a split culture regime for cell maintenance, the maximum attainable cell density without loss of cell viability has to be assessed. Cells should be maintained within these limits to retain maximum viability. Both suspension and anchorage-dependent cell types can be maintained as static culture in standard tissue culture flasks or medical flat bottles which have been treated for animal cell growth. However, if the suspension cells are specifically for fermentation in STRs or air-lift reactors better results are obtained if they are maintained in a suspension mode. The most appropriate way of achieving this is by employing magnetically agitated flasks such as Belco or Techne spinner flasks or round-bottomed flasks which provide a large surface area for surface gassing. The magnetic stirrer employed should not generate heat. Systems are commercially available which meet these requirements. Cells can be maintained in a viable condition by employ-

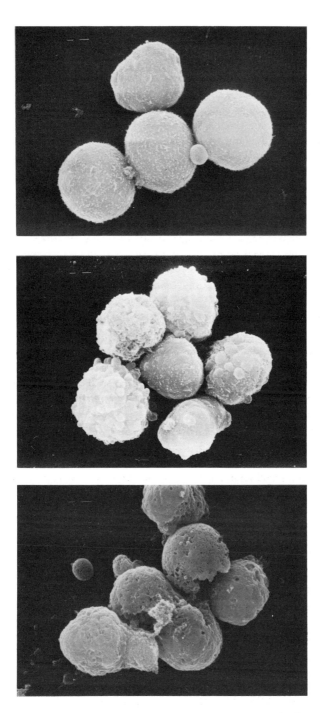

Figure 4. Scanning electron microscope pictures of cells at various stages of culture. Top, at day 1 (magnification 3.8K); middle, at day 3 (magnification × 3.4K); bottom, at day 5 (magnification × 3.9K).

ing the split culture regime on the agitated cells, from which a seed culture can be raised for inoculation of the main fermenter. *Figure 4* illustrates the appearance of the cells during cultivation.

One of the major problems in animal cell culture is contamination or infection. Contamination can be avoided by the application of good hygiene and good laboratory practice. Antibiotics such as penicillin/streptomycin combinations are also commonly added to cell culture medium to reduce the risk of contamination. All manipulations with animal cell cultures should be carried out in a standard laminar flow cabinet fitted with Hepa-type filters. It is also a valuable and worthwhile investment to enclose the fermenter vessels in a sterile laminar flow environment; this will reduce risks of contamination, particularly on seed transfer to vessel and medium supplement additions. In contrast to most microbial cells, animal cells are rather slow growing and therefore bacterial or fungal contamination in a culture can usually be easily seen. Infection of cells, however, usually occurs as a result of slow-growing microbes, the most common being mycoplasmas. This is more difficult to detect visually and requires selective media or special staining techniques. Tissue culture medium manufacturers such as Flow and Gibco, in addition to supplying mycoplasma test and stain kits, also provide a mycoplasma detection service. This can prove costly and in some cases inconclusive; therefore, when contamination or infection is suspected it is better to discard the suspect cultures immediately, and if necessary return to stocks stored in liquid nitrogen.

8. TISSUE CULTURE MEDIA

Basal animal cell culture medium generally consists of a carbohydrate source such as glucose, fructose or galactose, a mixture of essential and non-essential amino acids, vitamins, inorganic salts and some trace elements. Traditional cell culture medium such as MEM, DMEM, RPMI 1640 and Hams' F12 are still widely used. An example of a typical culture medium is given in *Table 2*. Commercial supplies of these media can be obtained either as pre-sterilized $1 \times$ or $10 \times$ strength liquid or in the form of a powder. A desired medium formulation can of course be constructed from the individual components. However, the benefits obtained from using commercial supplies on the small scale outweigh any benefits which can be obtained from laboratory formulation.

If powdered medium is preferred, an ultrapure water supply system such as that supplied by Elga or Millipore and a $0.1 \mu m$ filtration system will be required. Although autoclavable tissue culture media are commercially available (Auto POW), the majority require membrane filtration to assure medium sterility. For fermentation work in STRs or air-lift reactors, the vessels have to be sterilized separately from the medium. There are several ways to tackle this problem.

(i)   A sufficient volume of water can be sterilized in the vessel and the desired quantity of $10 \times$ strength medium and supplements added aseptically.

(ii)   The vessel can be sterilized empty and sterile $1 \times$ strength medium with supplements pumped in through sterile connnections.

(iii)   The medium can be made from a powdered formulation and filter-sterilized into a pre-sterilized empty vessel.

Whichever method is chosen, the vessel should be left at the desired running temperature and with the appropriate gas mixture overnight to act as a sterility check and allow the medium to equilibrate. The seed can then be fed into the vessel either by peristaltic pump, pressure or gravity to initiate the fermentation.

Supplements which are added to the tissue culture medium include antibiotics and serum. Ideally the use of antibiotics should be avoided since resistant microbes may develop. It also tends to cover up sloppy technique. However, if they must be used, penicillin (100 IU/ml), streptomycin (50 μg/ml) or gentamicin (50 μg/ml) can be included as antibacterial agents and amphotericin B (25 μg/ml) or nystatin (25 μg/ml) as antifungal agents are the most commonly used.

Serum is used primarily in tissue culture medium as a non-defined source of essential nutrients. Although expensive, foetal calf serum (FCS) has become the most frequently used supplement, at a concentration of 0.5−30% v/v. Other sources of serum such as equine, newborn calf and porcine are cheaper and have also been used. FCS supplementation has advantages for *in vitro* monoclonal antibody production since it is generally very low in immunoglobulins. Another important function of serum in animal cell fermentation is its protective effect against possible damage during culture in STRs or air-lift reactors. Polymers such as pluronic F68 (a copolymer of polyoxyethylene and polyoxypropylene) and methylcellulose have been used at 1 g litre^{-1} as protective agents against mechanical damage. This approach may be required if low serum concentrations are desired.

The popular trend today is to replace serum with other non-defined low-cost supplements such as milk, peptone, lactalbumin, hydrolysate, yeast extract and egg yolk emulsion, or defined supplements such as combinations of albumin, transferrin and insulin. Albumin is probaby one of the less defined serum replacements and, depending upon the type employed, a source of fatty acids or lipids may also have to be used. A typical serum-free medium may consist of Iscove's basal medium supplemented with bovine serum albumin (500 mg litre^{-1}), bovine insulin (2.5 mg litre^{-1}) and transferrin (5 mg litre^{-1}). A variety of serum-free medium supplements are available commercially, although their use does not necessarily make the medium cheaper than FCS supplementation. It does, however, provide a more defined environment for fundamental research on cell physiology in chemostat culture and in some instances may facilitate purification of secreted products. One of the major advantages, however, is the absence of adventitious agents such as viruses, bacteriophage and mycoplasma. The use of serum-free medium also reduces the risk of medium variation caused by different batches of whole serum.

9. FURTHER READING

General

Pirt,S.J. (1975) *Principles of Microbe and Cell Cultivation*. Blackwell, Oxford.
Paul,J. (1975) *Cell and Tissue Culture* (5th edn). Livingston, Edinburgh and London.
Freshney,R.I. (1986) *Animal Cell Culture—A Practical Approach*. IRL Press, Oxford.

Microcarrier

Hirstenstein,M., Clark,J., Lindgren,G. and Vretblad,P. (1980) *Dev. Biol. Standard,* **46**, 109.

Van Wezel,A.L. (1985) In *Animal Cell Biotechnology.* Academic Press, New York, Vol. 1, p. 265.
Butler,M., Hassell,T. and Rowley,A. (1987) In *Process Possibilities* (Webb and Mavituna, eds), p. 64.

Culture media

Lambert,K.J. and Birch,J.R. (1985) In *Animal Cell Biotechnology.* Academic Press, New York, Vol. 1, p. 85.
De-Tymowski,A. (1985) *Biofutur,* **34**, 35.

Process intensification

Lambe,C.A. and Walker,A.G. (1987) *Plant and Animal Cell Cultures. Process Possibilities.* (Webb and Mavituna, eds), p. 117.
Tyo,M.A. and Spier,R.E. (1987) *Enzyme and Micro Technol.,* **9**, 514.

APPENDIX

Suppliers of specialist items

A. Culture collections

American Type Culture Collection, 12301 Parklawn Drive, Rockville, Maryland 20852, USA.

Centraalbureau voor Schimmelcultures (Fungi), Oosterstraat 1, PO Box 273, 3740 AG Baarn, The Netherlands.

Culture Centre of Algae and Protozoa, 36 Storey's Way, Cambridge CB3 0DT, UK.

Culture Collection, Commonwealth Mycological Institute (Fungi), Ferry Lane, Kew TW9 3AF, UK.

National Collection of Industrial Bacteria, Torry Research Station, PO Box 31, 135 Abbey Road, Aberdeen AB9 8DG, UK.

National Collection of Type Cultures, Central Public Health Laboratory, Colindale Avenue, London NW9 5HT, UK.

National Collection of Yeast Cultures, Food Research Institute, Colney Lane, Norwich NR4 7UA, UK.

B. Fermentation equipment suppliers

Anglican Instruments, Newhaven BN9 9QJ, UK.

Alfa Laval (CHEMAP), Brentford TW9 3DT, UK.

APV Ltd, Hanley, Stoke-on-Trent ST1 4DW, UK.

Belco Biotechnology, London NW6 2BP, UK.

Bioengineering (UK) Ltd, Caterham, Surrey CR3 5UA, UK.

Bioreactor Technology Inc., Troy, New York 12180, USA.

B.Braun, Aylesbury HP20 1DQ, UK.

Camlab Ltd, Cambridge CB4 1TH, UK.

Fisons Ltd, Ipswich, UK.

F.T.Scientific Ltd, Bredon, Tewkesbury GL20 7HH, UK.

Gallenkamp, Loughborough LE11 0RG, UK.

L.H.Fermentation Ltd, Slough SL2 4EG, UK.

Life Science Laboratories, Luton LU4 9DT, UK.

MBR Bioreactor AG, Werkstrasse 4, CH-8620 Wetzikon, Switzerland. (Currently no UK agent.)

Millipore Ltd, Middlesex, UK.

New Brunswick Scientific, Watford, Herts WD2 4PT, UK.

Northern Media Supply Co. Ltd, Hessle, UK.

Pall Filtration Ltd, Portsmouth, UK.

Phase Separations Ltd, Queensferry, UK.

Sartorious (UK), Epsom, Surrey KT19 9QN, UK.

SGI (Setric), Newhaven BN9 0JX, UK.
Techne Ltd, Duxford, Cambridge CB2 4PZ, UK.
Watson-Marlow, Falmouth, Cornwall TR11 4RU, UK.

C. Probe electrode manufacturers and suppliers

Dr W.Ingold AG, Industrie Nord, CH-8902 Zurich, Switzerland. (UK agents are LSL.)
Kent Industrial Measurement Ltd, Stonehouse, Gloucestershire GL10 3TA, UK.
Life Science Laboratories Ltd, Luton LU4 9DT, UK.
New Brunswick Scientific Ltd, Watford WD2 4PT, UK.
Russell pH Ltd, Auchtermuchty, Fife KY14 7DP, UK.
Uniprobe Instruments Ltd, Cardiff CF5 1HG, UK.

D. Control equipment

Alfa Laval (see CHEMAP).
B.Braun Melsungen AG, Biotechnics Division, Melsungen, FRG.
Chemap AG, CH-8604 Volketswil, Switzerland.
BTG-EurControl Marketing SA, CH-1001 Lausanne, Switzerland.
Rintekno OY (see B.Braun).
Sartorius GmbH, Göttingen, FRG.

E. Hollow fibres

Amicon Ltd, Stonehouse, Gloucestershire GL10 2BJ, UK.
EDT Analytical Ltd, London NW10 7LU, UK.
Technogen Systems Ltd, South Harrow, Middlesex HA2 0LQ, UK.
Northumbria Biologicals Ltd, South Nelson Industrial Estate, Northumberland
NE23 9HL, UK.

INDEX

95657601
16/2/90